What are GI and GL?

Glycaemic Index

The glycaemic index (GI) is a measure of the ability of a food to raise blood glucose (sugar) levels after it has been eaten. The GI of a food is determined by several factors such as how much it's been processed, what type of cooking method has been used and the presence of nutrients such as protein and fat.

How does it work?

When glucose is released into the bloodstream during digestion, our body produces a hormone called insulin whose function is to transport glucose to places where it is needed for fuel – usually the muscle cells and the brain.

While our bodies can handle a steady release of glucose during digestion, when large quantities flood the bloodstream, the body's regulatory system goes into overload and produces large amounts of insulin to clear the glucose away into the body cells. This surge in insulin has the unfortunate effect of heightening our feelings of hunger which in turn increases our desire to eat more carbs.

Regrettably excess carbs get stored away in our fat cells. If this happens frequently, it can lead to weight gain and over time can even damage our cells by causing insulin resistance in which cells that normally respond to insulin become less sensitive to its effects.

Insulin resistance, high insulin levels and excess high GI foods have all been linked with Type II or middle-aged onset diabetes as well as premature ageing, heart disease and some cancers.

The benefits of eating low GI foods

Health experts believe that eating foods which cause a slow and steady rise of blood sugar levels, known as low GI foods, can have advantages for those watching their waistlines as well as for diabetic patients. This is because low GI foods help to delay feelings of hunger, compared to high GI foods thus helping you to control your calorie intake.

There are a number of scientific studies that show that people who eat low GI foods lose more body fat than those who eat high GI foods.

Glycaemic Load

The Glycaemic Load (GL) is a related measure and is calculated by multiplying the GI of a food by the amount of carbohydrate per typical serving and dividing it by 100.

GL is more reliable than the glycaemic index as a predictor of how a food will affect the blood sugar level. This is because some foods which have a high GI like carrots or watermelon contain such a small amount of carbohydrate in a normal portion that they would not be expected to raise the blood sugar levels very much. On the other hand, watermelon or carrot juice contains a larger amount of carbohydrate, which would cause a larger increase in blood sugar levels.

Glycaemic Load can be calculated using the following equation:

$$GL = \frac{GI \times \text{carbohydrate per portion}}{100}$$

New evidence is emerging that links high GL meals with an increased risk for heart disease and diabetes, especially in overweight and insulin-resistant people.

Understanding GI and GL ratings

- **Low GI and low GL** A food that has a low glycaemic load will have a small effect on blood glucose levels, as it either doesn't contain a lot of carbohydrate and/or it has a low GI score.
- **Low GI and high GL** Even low GI foods, if eaten in large quantities, can affect (raise) blood glucose levels quite significantly, especially if they are concentrated sources of carbohydrates (for example, most cakes, dried fruit and dried fruit bars, fruit juices and crisps).
- **High GI and low GL** These foods contain a small amount of carbohydrate in typical serving sizes and will not affect blood glucose levels significantly even though they have a high GI. It is important to remember that these should not be eaten with other high GI or high GL foods.
- **High GI and high GL** These foods will cause a dramatic increase in blood glucose levels and so should be avoided wherever possible.

Healthy eating

Whether you want to lose just a few kilograms or rather more, eating healthily is vital. This does not mean cutting out all of your favourite treats completely, just eating them in moderation. If you're not enjoying your food, you are less likely to stick to your diet and exercise regime.

Losing weight doesn't just mean that you'll have more energy and will feel better about yourself, it also decreases the risk of developing various ailments, including diabetes, hypertension, coronary heart disease, stroke, respiratory problems, gallstones and some cancers.

A balanced diet

Including foods from the five food groups means that you are meeting your requirements for nutrients.

THE FIVE GROUPS ARE:

1 Bread, cereals and potatoes This group is rich in starchy carbohydrates and includes breakfast cereals, rice, pasta, noodles, yams and oats and should form the basis of most meals. Foods in this group are rich in insoluble fibre, calcium, iron and B vitamins, which are needed to keep your gut, bones and blood healthy. Try to eat wholegrain, wholemeal or high-fibre versions of breads and cereals.

2 Fruits and vegetables These are important sources of antioxidants such as vitamin C and beta-carotene (vegetable vitamin A), which protect us from cancers and heart disease. They are also rich in soluble fibre, which helps lower blood cholesterol. Try to include five portions of different fruits and vegetables in your diet each day, whether fresh, frozen, canned, dried or juiced.

3 Milk and dairy foods These are excellent sources of calcium, protein and vitamins A and B12, and are essential for maintaining the health of bones, skin and blood. Include a couple of reduced-fat servings from this food group each day.

4 Meat, fish and alternatives The main nutrients supplied by this food group include iron, protein, B vitamins, zinc and magnesium, which help to maintain healthy blood and an efficient immune system. Daily, choose at most two servings of lean red meat, fish, chicken, nuts, turkey, eggs, beans or pulses. The last two are great protein alternatives, as is tofu, which is also a good source of calcium.

5 Foods containing sugar or fat Minimize your intake of savoury snacks, biscuits, cakes, crisps, pastries, sweets, chocolate, pies, butter and carbonated drinks, as these will hinder your efforts to lose weight.

How to use this book

This book lists the GI and GL rating, energy – as kilocalories (kcal) and kilojoules (kJ) – fat, saturated fat, protein, carbohydrate and fibre contained in more than 1,500 foods. Nutrient values have been expressed as average servings so no calculator is needed.

Guideline Daily Amounts

Guideline Daily Amounts (GDAs), developed by the Institute of Grocery Distribution, are a guide to how many calories, and how much sugar, fat, saturated fat and salt adult women and men should be eating everyday. These are average figures and personal requirements vary with age, weight and levels of activity

AVERAGE DAILY REQUIREMENT

kcal	2000
sugar	90g
fat	70g
saturated fat	20g
salt	6g

The GDA label gives how many calories and how much sugar, fat, saturated fat and salt a product contains and the percentage contribution to an adults' GDA for each of these nutrients.

Sometimes you will find that food products contain both traffic light information and GDA.

Highs and lows

For products that don't have front-of-pack traffic lights or GDAs, look at the nutritional information panel on the back of the food packaging and choose products by following these guidelines from the Food Standards Agency.

GUIDELINES PER 100G (3½OZ)

nutrient	high	low
sugar	15g or more	5g or less
fat	20g or more	3g or less
saturated fat	5g or more	1.5g or less
salt	1.5g or more	0.3g or less
sodium	0.4g or more	0.1g or less

Aim to include foods that are high in fibre in your diet too. Anything with more than 3g of fibre per 100g (3½oz) is a good choice.

Nutritional information

This book will provide you with at-a-glance tables of the calories, fat, saturated fat, protein, carbohydrate and fibre content per average portion of commonly available foods in the UK.

GI and GL ratings

Each food has been assigned a low, medium or high GI and a low, medium or high GL rating. The ratings are:

H = high
M = medium
L = low

FRUITS	AVERAGE PORTION g	GI	GL
Apples, cooking	130	L	L
Apples, cooking, stewed with sugar	110	M	M
Apples, cooking, stewed without sugar	140	L	L
Apples, eating	100	L	L
Apricots	80	M	M
Apricots, canned in juice	140	M	M
Apricots, dried	120	L	L
Apricots, stewed with sugar	140	M	M
Apricots, stewed without sugar	140	L	L
Avocado	145	L	L
Banana chips	13	M	H
Bananas	100	M	M
Bilberries	40	L	L
Blackberries	100	L	L
Blackberries, stewed with sugar	140	M	M
Blackberries, stewed without sugar	140	L	L
Blackcurrants	100	L	L
Blackcurrants, canned in juice	140	M	M
Blackcurrants, canned in syrup	140	M	M
Blackcurrants, stewed with sugar	140	M	M
Blackcurrants, stewed without sugar	140	L	L
Blueberries	100	L	L
Carambola	120	L	L
Cherries	80	L	L
Cherries, canned in syrup	68	M	M
Cherries, glacé	15	M	M
Clementines	60	L	L
Cranberries	75	L	L
Currants	25	H	H
Custard apples	60	L	L
Damsons	80	L	L
Damsons, stewed with sugar	100	M	M
Damsons, stewed without sugar	100	L	L
Dates	100	H	H

Unless otherwise stated, fruits are prepared but uncooked.

ENERGY kcal	ENERGY kJ	FAT g	SATURATED FAT g	PROTEIN g	CARBO-HYDRATE g	FIBRE g
46	196	Trace	Trace	0	12	2.1
81	345	Trace	Trace	0	21	1.3
56	237	0.4	0.1	0.3	13.6	1.8
51	215	0.5	0.1	0.6	11.6	1.3
25	107	Trace	Trace	1	6	1.4
48	206	Trace	Trace	1	12	1.3
190	809	1	Trace	5	44	7.6
101	431	Trace	Trace	1	26	2.2
38	161	Trace	Trace	1	9	2.
276	1137	28.3	5.9	2.8	2.8	4.9
66	278	4	Trace	0	8	0.2
81	348	0.1	0	1.2	20.3	0.8
12	51	Trace	Trace	0	3	0.7
25	104	Trace	Trace	1	5	3.1
78	335	Trace	Trace	1	19	3.4
29	123	Trace	Trace	1	6	3.6
28	121	Trace	Trace	1	7	3.6
43	189	Trace	Trace	1	11	4.3
101	428	Trace	Trace	1	26	3.6
81	353	Trace	Trace	1	21	3.9
34	144	Trace	Trace	1	8	4.3
40	169	0.2	0	0.9	9.1	1.5
38	163	1	0.1	1	9	1.6
38	162	Trace	Trace	1	9	0.7
48	207	Trace	Trace	0	13	0.4
39	159	Trace	Trace	0	9	0
22	95	Trace	Trace	1	5	0.7
11	49	Trace	Trace	0	3	2.3
67	285	Trace	Trace	1	17	0.5
41	178	Trace	Trace	1	10	1.4
30	130	Trace	Trace	0	8	1.4
74	316	Trace	Trace	0	19	1.5
34	147	Trace	Trace	1	9	1.6
124	530	Trace	Trace	2	31	1.8

FRUITS	AVERAGE PORTION g	GI	GL
Dates, stoned and dried	60	H	H
Durian	80	L	L
Figs	55	L	L
Figs, dried	84	H	H
Fruit cocktail, canned in juice	115	L	M
Fruit cocktail, canned in syrup	115	M	L
Fruit salad, mixed	140	L	L
Goji berries	10	L	L
Gooseberries	100	L	L
Gooseberries, canned in syrup	140	M	M
Gooseberries, stewed with sugar	140	M	M
Gooseberries, stewed without sugar	140	L	L
Grapefruit	231	L	L
Grapefruit, canned in juice	120	L	L
Grapefruit, canned in syrup	120	M	M
Grapes, green	100	L	L
Grapes, red	100	L	L
Greengages	100	L	L
Greengages, stewed with sugar	100	M	M
Greengages, stewed without sugar	100	L	L
Guavas	100	L	L
Guavas, canned in syrup	113	M	M
Kiwi fruit	60	L	L
Kumquats	8	L	L
Kumquats, canned in syrup	8	M	M
Lemons, unpeeled	60	L	L
Limes, unpeeled	40	L	L
Loganberries, canned in juice	140	L	L
Loganberries, stewed with sugar	140	M	M
Loganberries, stewed without sugar	140	L	L
Lychees	90	L	L
Lychees, canned in syrup	80	M	H
Mandarin oranges, canned in juice	115	L	L
Mandarin oranges, canned in syrup	126	L	M

Unless otherwise stated, fruits are prepared but uncooked.

ENERGY kcal	ENERGY kJ	FAT g	SATURATED FAT g	PROTEIN g	CARBO-HYDRATE g	FIBRE g
162	691	1	0.1	2	41	2.4
109	460	1	Trace	2	23	3
24	102	1	0.1	1	5	0.8
176	747	1	Trace	3	41	5.8
33	140	Trace	Trace	0	8	1.2
66	281	Trace	Trace	0	17	1.2
77	332	Trace	Trace	1	19	2.1
33	138	0.2	0	1.4	6.9	0.5
40	170	1	0.1	1	9	2.4
102	434	Trace	Trace	1	26	2.4
76	321	Trace	Trace	1	18	2.7
22	92	Trace	Trace	1	4	2.8
69	291	Trace	Trace	2	16	3
36	144	Trace	Trace	1	9	0.5
72	308	Trace	Trace	1	19	0.7
62	263	0.2	0	0.7	15.2	0.7
67	286	0.1	0	0.6	17	0.6
41	173	Trace	Trace	1	10	2.1
81	347	Trace	Trace	1	21	1.9
36	155	Trace	Trace	1	9	1.9
26	112	1	Trace	1	5	3.7
68	292	Trace	Trace	0	18	3.4
29	124	Trace	Trace	1	6	1.1
3	15	Trace	Trace	0	1	0.3
11	46	Trace	Trace	0	3	0.1
8	36	1	0.1	0	1	1.7
4	14	0	0	0	0	1.1
141	601	Trace	Trace	1	37	2.2
70	300	Trace	Trace	1	18	2.8
20	87	Trace	Trace	1	4	2.9
52	223	Trace	Trace	1	13	0.6
54	232	Trace	Trace	0	14	0.4
37	155	Trace	Trace	1	9	0.3
66	281	Trace	Trace	1	17	0.3

FRUITS	AVERAGE PORTION g	GI	GL
Mangoes	150	L	L
Mangoes, canned in syrup	105	M	L
Mangosteen	60	L	L
Melon, Canteloupe	150	H	H
Melon, Galia	150	H	H
Melon, Honeydew	200	H	H
Mixed peel	5	H	L
Nectarines	150	L	L
Oranges	160	L	L
Papaya	140	M	L
Passion fruit	60	L	L
Peaches	110	L	L
Peaches, canned in juice	120	L	L
Peaches, canned in syrup	120	M	M
Pears, canned in juice	135	L	L
Pears, canned in syrup	135	M	M
Pears, Comice	150	L	L
Pears, Conference	170	M	M
Pears, Nashi	150	L	L
Pears, William	150	L	L
Physalis	60	L	L
Pineapple	80	M	L
Pineapple, canned in juice	40	L	L
Pineapple, canned in syrup	40	M	M
Plums, average, stewed with sugar	133	M	M
Plums, average, stewed without sugar	70	L	L
Plums, canned in syrup	80	M	M
Plums, Victoria	55	L	L
Plums, yellow	55	L	L
Pomegranate	55	L	L
Pomelo	80	L	L
Prickly pears	60	L	L
Prunes, canned in juice	24	L	L
Prunes, canned in syrup	24	L	L

Unless otherwise stated, fruits are prepared but uncooked.

ENERGY kcal	ENERGY kJ	FAT g	SATURATED FAT g	PROTEIN g	CARBO-HYDRATE g	FIBRE g
86	368	1	0.2	1	21	3.9
81	347	Trace	Trace	0	21	0.7
44	184	Trace	Trace	0	10	1
29	122	Trace	Trace	1	6	1.5
36	153	Trace	Trace	1	8	0.6
56	238	Trace	Trace	1	13	1.2
12	49	Trace	Trace	0	3	0.2
60	257	Trace	Trace	2	14	1.8
58	243	0.3	0	1.3	13.1	2.7
50	214	Trace	Trace	1	12	3.1
22	91	1	0.1	2	3	2
36	156	Trace	Trace	1	8	1.7
47	198	Trace	Trace	1	12	1
66	280	Trace	Trace	1	17	1.1
45	190	Trace	Trace	0	11	1.9
68	290	Trace	Trace	0	18	1.5
50	212	Trace	Trace	0	13	3
90	386	1	Trace	1	22	4.1
44	183	Trace	Trace	0	11	2.3
51	215	Trace	Trace	1	12	3.3
32	131	Trace	Trace	1	7	1
33	141	Trace	Trace	0	8	1
19	80	Trace	Trace	0	5	0.2
26	109	Trace	Trace	0	7	0.3
105	446	Trace	Trace	1	27	1.7
21	90	Trace	Trace	0	5	0.9
47	202	Trace	Trace	0	12	0.6
21	92	Trace	Trace	0	5	1
14	59	Trace	Trace	0	3	0.6
28	120	Trace	Trace	1	6	1.9
24	101	Trace	Trace	0	5	0.8
29	124	1	0.1	0	7	2.2
19	80	Trace	Trace	0	5	0.6
22	93	Trace	Trace	0	6	0.7

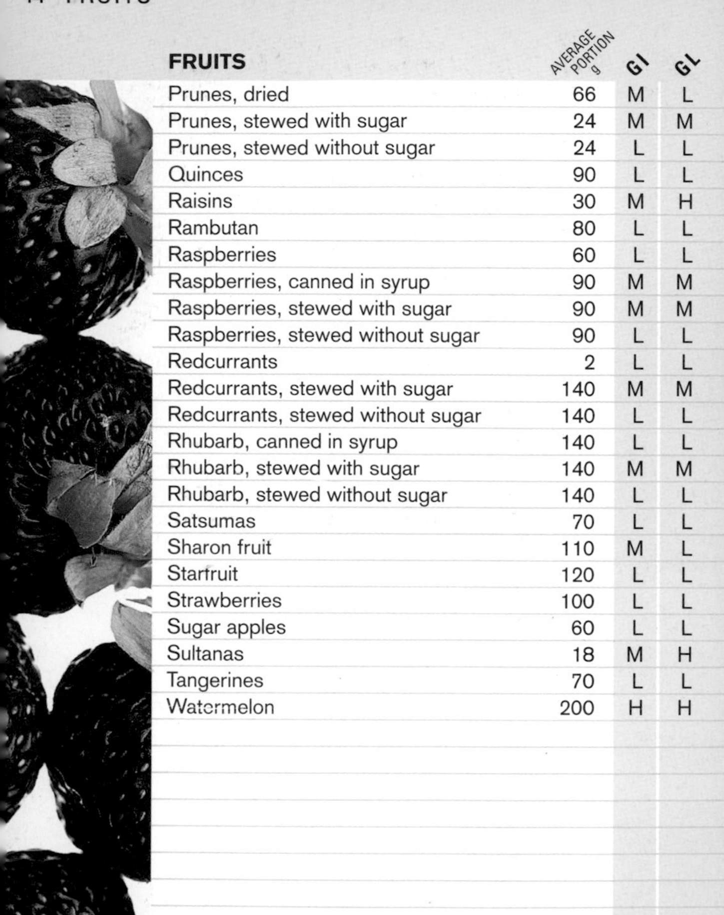

FRUITS	AVERAGE PORTION g	GI	GL
Prunes, dried	66	M	L
Prunes, stewed with sugar	24	M	M
Prunes, stewed without sugar	24	L	L
Quinces	90	L	L
Raisins	30	M	H
Rambutan	80	L	L
Raspberries	60	L	L
Raspberries, canned in syrup	90	M	M
Raspberries, stewed with sugar	90	M	M
Raspberries, stewed without sugar	90	L	L
Redcurrants	2	L	L
Redcurrants, stewed with sugar	140	M	M
Redcurrants, stewed without sugar	140	L	L
Rhubarb, canned in syrup	140	L	L
Rhubarb, stewed with sugar	140	M	M
Rhubarb, stewed without sugar	140	L	L
Satsumas	70	L	L
Sharon fruit	110	M	L
Starfruit	120	L	L
Strawberries	100	L	L
Sugar apples	60	L	L
Sultanas	18	M	H
Tangerines	70	L	L
Watermelon	200	H	H

Unless otherwise stated, fruits are prepared but uncooked.

ENERGY kcal	ENERGY kJ	FAT g	SATURATED FAT g	PROTEIN g	CARBO-HYDRATE g	FIBRE g
93	397	Trace	Trace	2	22	3.8
25	105	Trace	Trace	0	6	0.7
19	83	Trace	0	0	5	0.8
23	99	Trace	Trace	0	6	1.7
82	348	Trace	Trace	1	21	0.6
55	234	Trace	Trace	1	13	0.5
15	65	1	0.1	1	3	1.5
79	337	Trace	Trace	1	20	1.4
57	244	1	0.1	1	14	2
22	95	1	0.1	1	4	2.2
0	2	Trace	Trace	0	0	0.1
74	318	Trace	Trace	1	19	3.8
24	106	Trace	Trace	1	5	4.1
43	182	Trace	Trace	1	11	1.1
67	284	Trace	Trace	1	16	1.7
10	42	Trace	Trace	1	1	1.8
25	109	Trace	Trace	1	6	0.9
80	342	Trace	Trace	1	20	1.8
38	163	1	0.1	1	9	1.6
30	126	0.5	0.1	0.6	6.1	1
41	178	Trace	Trace	1	10	1.4
50	211	Trace	Trace	0	12	0.4
25	103	Trace	Trace	1	6	0.9
62	266	1	0.2	1	14	0.2

VEGETABLES	AVERAGE PORTION g	GI	GL
Ackee, canned	80	L	L
Alfalfa sprouts	5	L	L
Artichoke, globe, heart	40	L	L
Artichoke, Jerusalem	56	M	L
Asparagus	125	L	L
Aubergine, fried	130	L	L
Bamboo shoots, canned	50	L	L
Beans			
Aduki, dried, boiled	60	L	L
Baked, canned in tomato sauce	135	L	L
Baked, canned in tomato sauce, reduced sugar and salt	135	M	M
Baked, canned in tomato sauce, with burgers	225	M	M
Baked, canned in tomato sauce, with pork sausages	225	M	M
Balor, canned	135	L	L
Barbecue, canned in sauce	135	M	M
Blackeye, dried, boiled	60	L	M
Broad	120	H	L
Broad, canned	120	H	L
Broad, frozen	120	H	L
Butter, canned	120	L	L
Butter, dried, boiled	60	L	L
Chickpeas, canned	70	L	L
Chickpeas, split, dried, boiled	70	L	L
Chickpeas, whole, dried, boiled	70	L	L
Chilli, canned	135	L	L
Edamame (soya) beans	25	L	L
French	90	L	L
French, canned	90	L	L
French, frozen	90	L	L
Green, boiled	90	L	L
Green, canned	90	L	L

Unless otherwise stated, vegetables are described as they would normally be eaten.

ENERGY kcal	ENERGY kJ	FAT g	SATURATED FAT g	PROTEIN g	CARBO-HYDRATE g	FIBRE g
121	500	12	Trace	2	1	1.4
1	5	Trace	Trace	0	0	0.1
7	31	0	0	1	1	2
23	116	Trace	Trace	1	6	2
33	138	1	0.1	4	2	1.7
393	1620	41	5.3	2	4	3
6	23	1	0.1	1	0	0.9
74	315	Trace	Trace	6	14	3.3
109	463	0.7	0.1	6.8	20.3	5.1
99	420	1	0.1	7	17	5.1
214	900	7	0.1	15	25	6.8
239	1001	10	0.1	14	25	6.3
26	112	Trace	Trace	3	4	3.6
104	444	1	0.1	7	19	4.7
70	296	1	0.1	5	12	2.1
58	245	1	0.1	6	7	6.5
104	444	1	0.1	10	15	6.2
97	413	1	0.1	9	14	7.8
92	392	1	0.1	7	16	5.5
62	262	1	0.1	4	11	3.1
81	341	2	0.2	5	11	2.9
80	339	1	0.1	5	12	3
85	358	1	0.1	6	13	3
95	513	1	0.1	7	16	5.3
111	464	5.2	0.8	10	4	3.7
20	83	1	0.1	2	3	2.2
20	86	Trace	Trace	1	4	2.3
23	97	Trace	Trace	2	4	3.7
23	98	0.3	1.8	1.9	3.6	2.3
20	86	Trace	Trace	1	4	2.3

VEGETABLES	AVERAGE PORTION g	GI	GL
Green, frozen	90	L	L
Green, raw	90	L	L
Haricot, dried, boiled	60	L	M
Lilva, canned	80	L	L
Mung, dahl, dried, boiled	60	L	L
Mung, whole, dried, boiled	60	L	L
Papri, canned	60	L	L
Pigeon peas, dahl, dried	60	L	L
Pinto, dried, boiled	60	L	L
Pinto, refried	60	L	L
Red kidney, canned	60	L	L
Red kidney, dried, boiled	60	L	L
Runner	90	L	L
Soya, dried, boiled	60	L	L
Beansprouts, mung	80	L	L
Beansprouts, mung, canned	80	L	L
Beansprouts, mung, stir-fried	80	L	L
Beetroot	40	M	L
Beetroot, pickled	35	M	L
Breadfruit, canned	40	L	L
Broccoli, green	85	L	L
Broccoli, purple sprouting	85	L	L
Brussels sprouts	90	L	L
Cabbage, Chinese	40	L	L
Cabbage, green	95	L	L
Cabbage, green, raw	90	L	L
Cabbage, red	90	L	L
Cabbage, red, raw	90	L	L
Cabbage, Savoy	95	L	L
Cabbage, white	95	L	L
Cabbage, white, raw	90	L	L
Carrots, canned	60	M	L
Carrots, old, boiled	60	M	L
Carrots, old, raw	80	M	L

Unless otherwise stated, vegetables are described as they would normally be eaten.

ENERGY kcal	ENERGY kJ	FAT g	SATURATED FAT g	PROTEIN g	CARBO-HYDRATE g	FIBRE g
23	97	Trace	Trace	2	4	3.7
21	89	0.4	2.9	1.9	2.8	2.3
57	244	1	0.1	4	10	3.7
54	232	1	0.1	5	8	0.5
55	235	1	0.1	5	9	1.8
55	233	1	0.1	5	9	1.8
16	66	1	0.1	2	2	0.4
71	298	1	0.1	5	13	4
82	350	1	0.1	5	14	2.8
64	268	1	0.1	4	9	3.2
60	254	1	0.1	4	11	3.7
62	264	1	0.1	5	10	4
16	68	1	0.1	1	2	1.7
85	354	4	0.5	8	3	3.7
25	105	1	0.1	2	3	1.2
8	35	Trace	Trace	1	1	0.6
58	238	5	0.4	2	2	0.7
18	78	Trace	Trace	1	4	0.8
10	41	Trace	Trace	0	2	0.6
26	112	Trace	Trace	0	7	0.7
20	85	1	0.2	3	1	2
16	68	1	0.1	2	1	2
32	138	1	0.3	3	3	2.8
5	20	Trace	Trace	0	1	0.5
16	67	0.2	0	1.4	2.2	2.5
24	103	0.2	0	2.2	3.7	2.4
14	55	Trace	Trace	1	2	1.8
19	80	Trace	Trace	1	3	2.3
16	67	1	0.1	1	2	1.9
13	57	Trace	Trace	1	2	1.4
22	91	0.1	0	1.1	4.3	1.8
12	52	1	0.1	0	3	1.1
66	275	0.3	0	0.3	2.9	1.3
27	117	0.3	0.1	0.4	6.2	1.7

VEGETABLES	AVERAGE PORTION g	GI	GL
Carrots, young	60	M	L
Carrots, young, raw	60	M	L
Cassava, baked	100	M	M
Cauliflower, boiled	90	L	L
Cauliflower, raw	90	L	L
Celeriac	30	L	L
Celeriac, raw	30	L	L
Celery	30	L	L
Chard, Swiss	90	L	L
Chicory, raw	28	L	L
Chilli peppers, capsicum red	10	L	L
Chinese leaf	40	L	L
Corn, on the cob	200	M	M
Courgettes	90	L	L
Courgettes, fried	90	L	L
Cress	5	L	L
Cucumber	25	L	L
Curly kale	95	L	L
Curly kale, boiled	100	L	L
Endive	30	L	L
Fennel, Florence	40	L	L
Garlic, raw	3	L	L
Gourd, karela, canned	30	L	L
Jackfruit, canned	60	L	L
Kohlrabi	60	L	L
Kohlrabi, cooked	90	L	L
Leeks	75	L	L
Lentils, canned in tomato sauce	80	M	L
Lentils, green or brown, whole, dried	80	L	L
Lentils, red, split, dried, boiled	80	L	L
Lettuce	80	L	L
Marrow	65	L	L
Mixed vegetables, frozen	90	L	L
Mushrooms, cooked in sunflower oil	44	L	L

Unless otherwise stated, vegetables are described as they would normally be eaten.

ENERGY kcal	ENERGY kJ	FAT g	SATURATED FAT g	PROTEIN g	CARBO-HYDRATE g	FIBRE g
13	56	1	0.1	0	3	1.4
18	75	1	0.1	0	4	1.4
155	661	1	0.1	1	40	1.7
21	89	0.8	0.2	1.7	1.9	1.4
27	108	0.4	0	2.3	4	1.6
5	19	Trace	Trace	0	1	1
5	22	Trace	Trace	0	1	1.1
2	10	Trace	Trace	0	0	0.3
18	76	Trace	Trace	2	3	1.8
3	13	1	0.1	0	1	0.3
3	11	Trace	Trace	0	0	0.1
5	20	Trace	Trace	0	1	0.5
132	560	3	0.4	5	23	2.6
17	73	1	0.1	2	2	1.1
57	239	4	0.5	2	2	1.1
1	3	Trace	Trace	0	0	0.1
3	15	0.2	0	0.3	0.3	0.2
23	95	1	0.2	2	1	2.7
24	100	1.1	0.2	2.4	1	2.8
4	16	0	0	1	0	0.6
5	20	Trace	Trace	0	1	1
3	12	0	0	0	0	0.1
3	11	Trace	Trace	0	0	0.7
62	264	Trace	Trace	0	16	0.9
14	57	Trace	Trace	1	2	1.3
16	69	Trace	Trace	1	3	1.7
16	65	1	0.1	1	2	1.3
44	189	Trace	Trace	4	7	1.4
84	357	1	0.1	7	14	3
80	339	Trace	Trace	6	14	1.5
12	49	0.4	0.1	1	1.1	1
6	25	Trace	Trace	0	1	0.4
38	162	Trace	Trace	3	6	2.9
48	202	4.8	0.9	1.1	0.1	0.7

VEGETABLES	AVERAGE PORTION g	GI	GL
Mushrooms, raw	40	L	L
Mushrooms, stewed	40	L	L
Okra	30	L	L
Onions, fried	40	L	L
Onions, raw	150	L	L
Pak choi, steamed	90	L	L
Parsley, fresh	10	L	L
Parsnips	65	H	M
Parsnips, roasted without oil	90	H	M
Peas, canned	70	M	L
Peas, fresh	70	M	L
Peas, frozen	70	L	L
Peas, frozen, microwaved	70	L	L
Peas, mangetout, stir-fried	90	L	L
Peas, marrowfat, canned	80	M	L
Peas, mushy, canned	80	M	L
Peas, petit pois, canned	70	M	L
Peas, petit pois, frozen	70	M	L
Peas, sugar-snap	90	L	L
Peppers, capsicum, green	160	L	L
Peppers, capsicum, red	100	L	L
Peppers, capsicum red, cooked	100	L	L
Peppers, capsicum, yellow	160	L	L
Plantain	200	L	M
Plantain, ripe, fried	200	L	M
Potatoes and potato products			
Chips, fine cut, takeway	165	H	H
Chips, oven baked	165	H	H
Chips, takeaway	165	H	H
Croquettes, fried	180	H	H
Duchesse	120	H	H
French fries	110	H	H
Instant mash, made with semi-skimmed milk	60	H	M

Unless otherwise stated, vegetables are described as they would normally be eaten.

ENERGY kcal	ENERGY kJ	FAT g	SATURATED FAT g	PROTEIN g	CARBO-HYDRATE g	FIBRE g
3	12	0.1	0	0.4	0.1	0.5
4	15	0.1	0	0.6	0	0.8
8	36	1	0.1	1	1	1.1
66	274	4	0.6	1	6	1.2
53	225	0.2	0	1.5	12	1.7
13	53	0.1	1.8	1.4	1.7	0.9
3	14	0.1	0	0.3	0.3	0.5
43	181	1	0.1	1	8	3.1
102	427	6	Trace	1	12	4.2
56	237	1	0.1	4	9	3.6
55	230	1	0.2	5	7	3.2
48	202	0.5	0.1	3.7	7.5	2.7
50	212	0.6	0.1	4	7.6	3.2
64	268	4	0.4	3	3	2.2
80	329	1	0.1	6	14	3.3
65	276	1	0.1	5	11	1.4
32	132	1	0.1	4	3	3
34	144	1	0.1	4	4	3.2
30	125	1	0.1	3	4	1.2
24	104	1	0.2	1	4	2.6
29	121	0.2	0.1	0.8	6.4	1
46	193	0.1	0	0.8	11.2	0.8
42	181	Trace	Trace	2	8	2.7
224	954	1	0.2	2	57	2.4
534	2252	1	2	3	95	4.6
479	2011	23.4	4.1	5.8	65.5	5.3
312	1320	8.1	1.3	5.3	58.2	4.5
353	1488	13.9	7.1	5.8	54.8	6.1
385	1607	24	3.1	7	39	2.3
148	622	6	3.6	4	20	1.4
308	1291	17	6.4	4	37	2.3
42	178	1	0.2	1	9	0.6

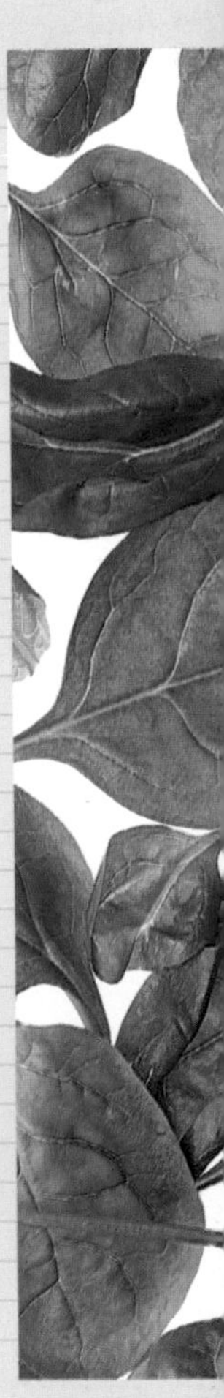

VEGETABLES	AVERAGE PORTION g	GI	GL
Instant mash, made with water	60	H	M
Instant mash, made with whole milk	60	H	M
New, canned	165	M	M
New, chipped, fried	165	H	H
New, in skins, boiled	175	H	M
Old, baked	180	H	M
Old, boiled	175	H	M
Old, mashed with butter	120	H	M
Old, mashed with polyunsaturated fat spread	120	H	M
Old, microwaved	180	H	M
Old, roast	130	H	M
Oven chips, frozen	165	H	H
Oven chips, thick cut, frozen	165	H	H
Shaped potato products, baked	100	H	H
Waffles, frozen	90	H	H
Wedges, baked	180	H	H
Pumpkin	60	H	L
Raddiccio, raw	30	L	L
Radishes, red, raw	48	L	L
Rocket leaves	20	L	L
Rosemary, fresh	10	L	L
Salad onions	10	L	L
Sauerkraut	30	L	L
Shallots	100	L	L
Spinach	90	L	L
Spinach, baby	90	L	L
Spinach, cooked	90	L	L
Split peas, dried, boiled	90	L	L
Spring greens, boiled	95	L	L
Spring onions	10	L	L
Squash, acorn, baked	65	L	L
Squash, butternut, baked	65	L	L

Unless otherwise stated, vegetables are described as they would normally be eaten.

ENERGY kcal	ENERGY kJ	FAT g	SATURATED FAT g	PROTEIN g	CARBO-HYDRATE g	FIBRE g
34	147	Trace	Trace	1	8	0.6
46	193	1	0.4	1	9	0.6
104	447	Trace	Trace	2	25	1.3
376	1582	16	2	7	55	2.8
119	509	1.1	0.2	3.2	26.1	1.6
175	743	0.4	0	4.5	40.7	2.5
130	551	0.2	0	3.2	30.6	1.8
125	526	5	3.4	2	19	1.3
125	526	5	1.1	2	19	1.3
166	706	0.2	0	4.7	38.7	2.5
194	819	6	0.8	4	34	2.3
267	1134	7	3	5	49	3.3
259	1096	7	2.8	5	46	3
190	799	8.3	0.9	2.5	28.1	2.8
180	758	12	1	2	28	2
245	1046	Trace	Trace	7	57	4.9
8	34	1	0.1	0	1	0.7
4	17	Trace	Trace	0	1	0.5
6	24	0	0	0	1	0.4
4	15	0.1	0	0.7	0	0.3
10	42	0.4	0	0.4	1.4	0.5
2	10	0	0	0	0	0.2
3	11	Trace	Trace	0	0	0.7
21	88	0.4	0	1.4	3.2	1.4
23	93	1	0.1	3	1	1.9
11	45	0.1	0	2.3	0.2	1.1
17	71	1	0.1	2	1	1.9
113	484	1	0.2	7	20	2.4
19	78	1	0.1	2	2	2.5
2	10	0	0	0	0	0.2
36	152	Trace	Trace	1	8	2.1
21	89	Trace	Trace	1	5	0.9

VEGETABLES	AVERAGE PORTION g	GI	GL
Squash, spaghetti, baked	65	L	L
Swede	60	H	L
Sweet potato	130	L	M
Sweet potato, baked	130	M	H
Sweetcorn, baby, canned	60	M	M
Sweetcorn kernels, boiled on cob	85	M	M
Sweetcorn, canned	85	M	M
Tarragon, fresh	10	L	L
Tomatoes, canned	200	L	L
Tomatoes, cherry	90	L	L
Tomatoes, grilled	85	L	L
Tomatoes, raw	85	L	L
Turnips	60	L	L
Water chestnuts, canned	28	L	L
Watercress	20	L	L
Yam	130	L	M
Yam, baked	130	M	M
Yam, steamed	130	M	M

PREPARED SALADS			
Baby leaf salad	50	L	L
Bean salad	200	L	L
Beetroot salad	100	M	L
Carrot and nut salad with French dressing	95	L	L
Coleslaw, economy, shop bought	100	L	L
Coleslaw, shop bought	100	L	L
Coleslaw, with mayonnaise	100	L	L
Coleslaw, with reduced-calorie dressing	100	L	L
Coleslaw, with vinaigrette	100	L	L
Florida salad	95	L	L
Greek salad	95	L	L

Unless otherwise stated, vegetables are described as they would normally be eaten.

ENERGY kcal	ENERGY kJ	FAT g	SATURATED FAT g	PROTEIN g	CARBO-HYDRATE g	FIBRE g
15	62	Trace	0.1	0	3	1.4
7	28	Trace	Trace	0	1	0.4
109	465	1	0.1	1	27	3
150	634	1	0.3	2	36	4.3
14	58	Trace	Trace	2	1	0.9
57	241	1.6	0.2	3.1	8.1	2.2
107	446	1.4	0.2	2.2	22.6	2.
5	21	0.1	0	0.3	0.6	0.5
38	160	0.2	0	2.2	7.6	1.4
18	74	0.2	0	1	3.2	1.1
14	57	0.1	0	0.5	2.9	0.9
14	57	0.3	0.1	0.4	2.6	0.9
7	31	1	Trace	0	1	1.1
9	37	Trace	Trace	0	3	2
4	19	1	0.1	1	0	0.3
173	738	1	0.1	2	43	1.8
190	846	1	0.1	3	49	2.2
148	634	1	0.1	2	37	1.7
9	40	Trace	Trace	Trace	2	0.7
294	1236	19	2	8	26	6
100	417	7	0.7	2	8	1.7
207	858	17	1.6	2	13	2.3
110	456	9.1	0.7	0.9	6.5	1.8
173	714	16.3	1.7	0.8	6	1.7
258	939	26	3.9	1	4	1.4
67	280	5	0.5	1	6	1.4
87	364	4	0.5	1	12	1.7
213	790	19	2.9	1	9	1
124	513	12	3.1	3	2	0.8

PREPARED SALADS

	AVERAGE PORTION g	GI	GL
Green salad	95	L	L
Herb salad	50	L	L
Pasta salad	95	L	M
Pasta salad, wholemeal	95	L	M
Potato salad, with mayonnaise	85	H	H
Potato salad, with reduced-calorie mayonnaise	85	H	H
Rice salad	90	H	H
Rice salad, brown	95	M	H
Tabbouleh	100	M	H
Tomato and onion salad	90	L	L
Waldorf salad	90	L	L

VEGETABLE DISHES, HOMEMADE

	AVERAGE PORTION g	GI	GL
Bean loaf	90	L	L
Bhaji, aubergine, pea, potato and cauliflower	70	M	H
Bhaji, potato and cauliflower, fried	70	M	H
Bhaji, potato and onion	70	M	H
Broccoli in cheese sauce, made with semi-skimmed milk	190	L	M
Broccoli in cheese sauce, made with whole milk	190	L	L
Bubble and squeak, fried	140	M	H
Cannelloni, spinach	340	M	H
Cannelloni, vegetable	340	M	H
Casserole, vegetable	220	M	H
Cauliflower cheese, made with semi-skimmed milk	200	L	M
Cauliflower cheese, made with whole milk	200	L	M
Cauliflower in white sauce, made with skimmed milk	200	L	M

Unless otherwise stated, vegetables are described as they would normally be eaten.

ENERGY kcal	ENERGY kJ	FAT g	SATURATED FAT g	PROTEIN g	CARBO-HYDRATE g	FIBRE g
11	48	0	Trace	1	2	1
9	40	Trace	Trace	Trace	Trace	0.7
121	473	7	1	2	13	1.5
124	493	7	1	3	13	2.6
203	757	18	2.6	1	10	0.8
82	349	3	0.3	1	13	0.7
149	628	7	1	3	21	0.6
159	668	7	1	3	23	1
119	496	5	0.4	3	17	1
65	272	5	0.6	1	4	0.9
174	662	16	2.1	1	7	1.2
134	564	7	0.7	6	14	3.9
49	209	2	0.2	2	7	2
214	888	15	1.8	5	14	2.7
112	468	7	4.6	1	12	1.1
211	880	14	7.2	12	9	2.9
224	939	16	8.2	12	9	2.9
174	727	13	1.4	2	14	2.1
449	1880	26	7.8	15	43	2.7
493	2067	31	11.6	15	43	2.4
114	486	1	0.2	5	23	4.6
200	840	13	6	12	10	2.6
210	880	14	6.6	12	10	2.6
122	512	6	2.2	7	10	2.2

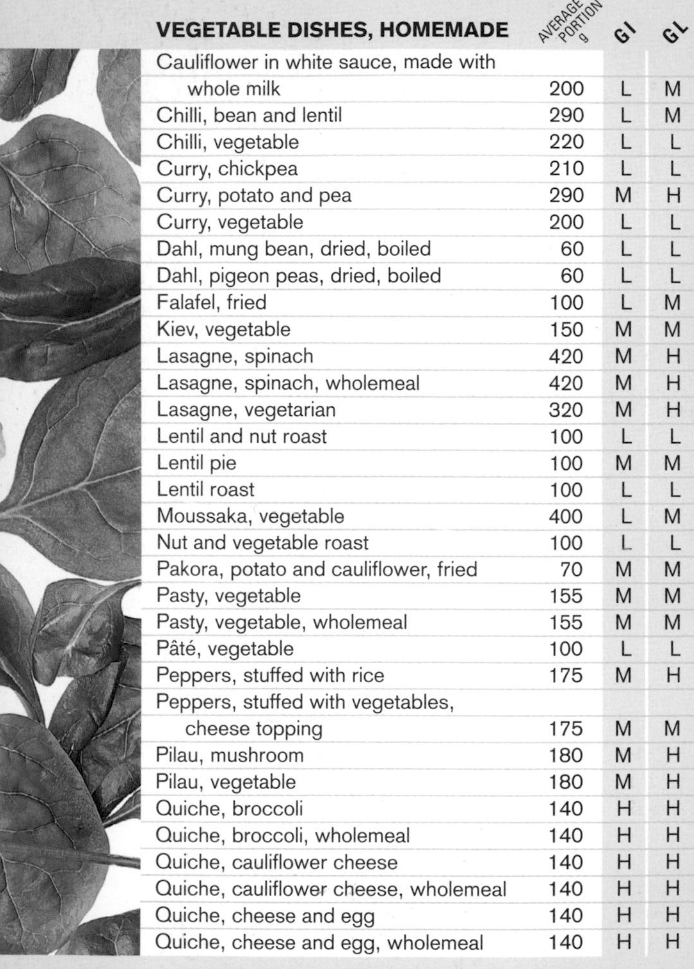

VEGETABLE DISHES, HOMEMADE	AVERAGE PORTION g	GI	GL
Cauliflower in white sauce, made with whole milk	200	L	M
Chilli, bean and lentil	290	L	M
Chilli, vegetable	220	L	L
Curry, chickpea	210	L	L
Curry, potato and pea	290	M	H
Curry, vegetable	200	L	L
Dahl, mung bean, dried, boiled	60	L	L
Dahl, pigeon peas, dried, boiled	60	L	L
Falafel, fried	100	L	M
Kiev, vegetable	150	M	M
Lasagne, spinach	420	M	H
Lasagne, spinach, wholemeal	420	M	H
Lasagne, vegetarian	320	M	H
Lentil and nut roast	100	L	L
Lentil pie	100	M	M
Lentil roast	100	L	L
Moussaka, vegetable	400	L	M
Nut and vegetable roast	100	L	L
Pakora, potato and cauliflower, fried	70	M	M
Pasty, vegetable	155	M	M
Pasty, vegetable, wholemeal	155	M	M
Pâté, vegetable	100	L	L
Peppers, stuffed with rice	175	M	H
Peppers, stuffed with vegetables, cheese topping	175	M	M
Pilau, mushroom	180	M	H
Pilau, vegetable	180	M	H
Quiche, broccoli	140	H	H
Quiche, broccoli, wholemeal	140	H	H
Quiche, cauliflower cheese	140	H	H
Quiche, cauliflower cheese, wholemeal	140	H	H
Quiche, cheese and egg	140	H	H
Quiche, cheese and egg, wholemeal	140	H	H

Unless otherwise stated, vegetables are described as they would normally be eaten.

ENERGY kcal	ENERGY kJ	FAT g	SATURATED FAT g	PROTEIN g	CARBO-HYDRATE g	FIBRE g
136	568	8	3.2	7	10	2.2
264	1111	8	0.9	15	38	10.4
125	532	1	0.2	7	24	5.7
227	956	8	0.8	13	30	6.9
267	1122	11	1.2	8	38	7
176	736	12	1.2	5	14	5
55	235	1	0.1	5	9	1.8
71	298	1	0.1	5	13	4
179	750	11	1.1	6	16	3.4
280	1172	16.6	6.2	6.5	26.7	5.7
365	1541	13	5.5	15	53	4.6
391	1659	13	5.5	18	55	9.7
458	1930	19.5	9	23.7	50.2	2.2
222	929	12	1.8	11	19	3.8
156	659	4	0.5	7	24	3.8
139	588	3	0.4	8	22	3.4
280	1172	10.8	5.2	10.8	36.4	9.6
314	1311	21.9	3.2	11.1	19.6	3.9
214	888	15	1.8	5	14	2.7
425	1783	23	5.7	6	52	2.9
406	1702	24	5.7	8	43	6.4
173	718	13.4	4.6	7.5	5.9	1.5
149	630	4	0.7	3	27	2.3
194	810	12	3.5	6	17	2.6
248	1048	8	4.5	4	43	0.7
248	1053	8	4.3	5	43	1
349	1455	21	8.3	12	30	1.7
337	1408	21	8.3	13	25	3.8
277	1156	18	7.1	7	24	1.5
269	1121	18	7.1	8	20	3.1
440	1834	31	14.4	18	24	0.8
431	1796	31	14.6	18	20	2.7

VEGETABLE DISHES, HOMEMADE	AVERAGE PORTION g	GI	GL
Quiche, cheese and mushroom	140	H	H
Quiche, cheese and mushroom, wholemeal	140	H	H
Quiche, cheese, onion and potato	140	H	H
Quiche, cheese, onion and potato, wholemeal	140	H	H
Quiche, mushroom	140	H	H
Quiche, mushroom, wholemeal	140	H	H
Quiche, spinach	140	H	H
Quiche, spinach, wholemeal	140	H	H
Quiche, vegetable	140	H	H
Quiche, vegetable, wholemeal	140	H	H
Refried beans	90	L	L
Rice and blackeye beans	200	M	H
Rice and blackeye beans, brown rice	200	L	H
Risotto, vegetable	290	M	H
Risotto, vegetable, brown rice	290	M	H
Rissoles, chickpea, fried	100	L	M
Rissoles, lentil, fried	100	L	M
Soup, broccoli and stilton	300	M	M
Soup, carrot and coriander	300	M	M
Soup, leek and potato	300	M	H
Soup, lentil	300	M	H
Soup, split pea	300	H	H
Soup, tomato, cream of	300	M	H
Vegetable bake	320	L	M
Vine leaves, stuffed with rice	80	L	M

Unless otherwise stated, vegetables are described as they would normally be eaten.

ENERGY kcal	ENERGY kJ	FAT g	SATURATED FAT g	PROTEIN g	CARBO-HYDRATE g	FIBRE g
396	1655	26	10.8	15	26	1.3
388	1616	27	10.8	16	22	3.1
480	2005	33	16	18	28	1.4
472	1966	34	16.1	19	25	3.1
398	1659	27	12.2	14	26	1.3
388	1618	28	12.2	15	21	3.1
287	1203	18	5.6	14	18	2
281	1176	18	5.6	15	16	3.2
295	1238	18	6	7	28	2.1
286	1197	18	6	8	24	3.9
211	885	12	2.7	9	18	6.3
366	1556	7	3	12	68	2.8
350	1488	7	3	11	66	3.6
426	1798	19	2.9	12	56	6.4
415	1749	19	2.6	12	54	7
243	1016	17	1.9	8	16	4.1
211	886	11	1.3	9	22	3.6
154	645	9.3	5	6.9	11.4	3.2
130	544	7.8	3	1.7	14.1	4.5
168	705	8.4	4.7	5.1	18.6	2.4
117	492	0.6	0	9.3	19.5	3.6
154	644	2.8	1.3	8.3	25.4	4.8
186	774	10.2	1.5	2.7	21.9	1.5
419	1754	23	9.3	13.8	41.9	3
210	875	14	2.1	2	19	1

VEGETARIAN PRODUCTS AND DISHES

	AVERAGE PORTION g	GI	GL
Beanburger	100	M	M
QuornTM, myco-protein	90	L	L
Quorn™ burger	50	L	L
Quorn™ mince	100	L	L
Quorn™ nuggets	100	L	L
Quorn™ roast	100	L	L
Quorn™ sausage	50	L	L
Soy burger	100	L	L
Soy mince	100	L	L
Tempeh	60	L	L
Tofu, fried	80	L	L
Tofu, steamed	80	L	L
Tofu, steamed, fried	80	L	L
Tofu burger, baked	9	L	L
Vegebanger mix, made with water, fried	56	M	M
Vegebanger mix, made with water and egg, fried	56	L	L
Vegeburger, fried	56	L	L
Vegeburger, grilled	56	L	L
Vegeburger mix, made with water, fried	56	M	M
Vegeburger mix, made with water, grilled	56	M	M
Vegeburger mix, made with water and egg, fried	56	M	L
Vegeburger mix, made with water and egg, grilled	56	M	L

Unless otherwise stated, vegetables are described as they would normally be eaten.

ENERGY kcal	ENERGY kJ	FAT g	SATURATED FAT g	PROTEIN g	CARBO-HYDRATE g	FIBRE g
190	795	7.6	0.6	9.6	20.7	4.6
77	326	3	Trace	11	2	4.3
80	321	3.7	1.3	7.1	4.1	1.2
94	397	2	0.5	14.5	4.5	5.5
182	761	10.5	1.5	12	9.9	4
106	445	2	0.5	15	4.5	4.9
86	362	3.5	0.3	6.8	6	1.8
179	749	6	0.7	17.9	13.4	4.6
303	1270	18.7	1.5	29.1	5	1.7
116	485	6.5	1.3	11.1	5.6	2.6
242	1011	20	2.3	17	3	1
58	243	3	0.4	6	1	1
209	869	14	2.3	19	2	1
11	45	0	0.1	1	1	0.2
137	572	9	2	8	7	1.8
157	659	10	2.5	10	7	1.9
137	571	10	1.7	9	4	2
110	460	6	1.7	9	4	2.4
115	482	6	1.3	7	8	2
84	356	3	1	7	8	2
122	511	7	1.6	8	7	1.8
91	385	4	1.2	8	7	1.8

BEEF	AVERAGE PORTION g	GI	GL
Braising steak, braised	140	L	L
Braising steak, untrimmed, braised	140	L	L
Braising steak, slow cooked	140	L	L
Braising steak, untrimmed, slow cooked	140	L	L
Fillet steak, fried	172	L	L
Fillet steak, untrimmed, fried	172	L	L
Fillet steak, grilled	168	L	L
Fillet steak, untrimmed, grilled	168	L	L
Flank, pot roasted	140	L	L
Flank, untrimmed, pot roasted	140	L	L
Fore rib, roasted	90	L	L
Fore rib, untrimmed, roasted	90	L	L
Meatballs, cooked	100	L	L
Mince, extra lean, stewed	140	L	L
Mince, microwaved	140	L	L
Mince, stewed	140	L	L
Rib eye steak, grilled	100	L	L
Rib roast, roasted	90	L	L
Rib roast, untrimmed, roasted	90	L	L
Rump steak, grilled	163	L	L
Rump steak, untrimmed, grilled	163	L	L
Rump steak, fried	166	L	L
Rump steak, untrimmed, fried	166	L	L
Rump steak strips, stir-fried	103	L	L
Rump steak strips, untrimmed, stir-fried	103	L	L
Silverside, pot roasted	140	L	L
Silverside, untrimmed, pot roasted	140	L	L
Silverside, salted, boiled	140	L	L
Silverside, untrimmed, salted, boiled	140	L	L
Sirloin joint, roasted	90	L	L
Sirloin joint, untrimmed, roasted	90	L	L
Sirloin steak, fried	169	L	L
Sirloin steak, untrimmed, fried	169	L	L
Sirloin steak, grilled medium rare	166	L	L

Unless otherwise stated, all meat is lean and trimmed, and all chops and cutlets are boned. Steaks are medium-sized. Unless otherwise stated, all dishes are homemade.

ENERGY kcal	ENERGY kJ	FAT g	SATURATED FAT g	PROTEIN g	CARBO-HYDRATE g	FIBRE g
315	1322	14	5.7	48	0	0
344	1441	18	7.4	46	0	0
276	1156	11	4.8	44	0	0
304	1270	16	6.7	41	0	0
316	1328	14	5.8	49	0	0
330	1384	15	6.7	49	0	0
316	1329	13	6	49	0	0
336	1410	16	7.4	48	0	0
354	1483	20	8	45	0	0
433	1800	31	12.7	38	0	0
212	889	10	4.6	30	0	0
270	1125	18	8.3	26	0	0
195	814	13.2	5	12.4	7.6	0.5
248	1039	12	5.3	35	0	0
368	1534	25	10.6	37	0	0
293	1218	19	8.3	31	0	0
155	649	8.3	3.2	20.1	0	0
212	889	10	4.6	30	0	0
270	1125	18	8.3	26	0	0
287	1208	9	3.9	51	0	0
331	1384	15	6.5	48	0	0
304	1278	11	4.2	51	0	0
378	1582	21	8.1	47	0	0
214	901	9	3.4	33	0	0
255	1069	15	5.8	31	0	0
270	1135	9	3.5	48	0	0
346	1448	19	7.8	43	0	0
258	1081	10	3.5	43	0	0
314	1312	18	6.6	39	0	0
169	712	6	2.6	29	0	0
210	876	11	5	27	0	0
319	1340	14	5.7	49	0	0
394	1646	24	10.1	45	0	0
292	1223	13	5.6	44	0	0

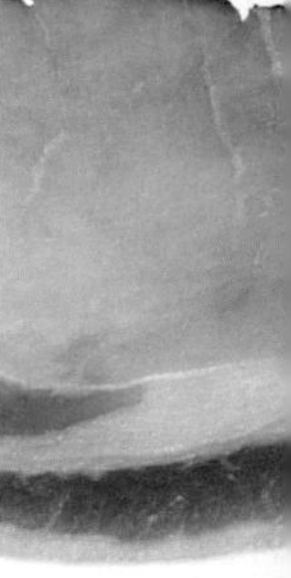

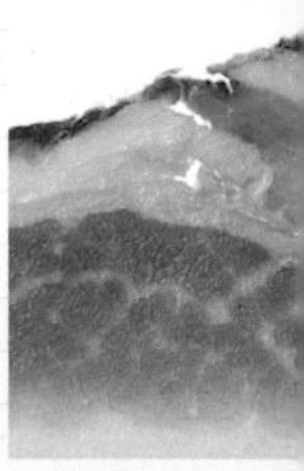

BEEF	AVERAGE PORTION g	GI	GL
Sirloin steak, untrimmed, grilled medium rare	166	L	L
Sirloin steak, grilled rare	166	L	L
Sirloin steak, untrimmed, grilled rare	166	L	L
Sirloin steak, grilled well done	166	L	L
Sirloin steak, untrimmed, grilled well done	166	L	L
Stewing steak, stewed	140	L	L
Stewing steak, untrimmed, stewed	140	L	L
Topside, roasted medium rare	90	L	L
Topside, untrimmed, roasted medium rare	90	L	L
Topside, roasted well done	90	L	L
Topside, untrimmed, roasted well done	90	L	L

LAMB			
Best end neck cutlets, grilled	50	L	L
Best end neck cutlets, untrimmed, grilled	50	L	L
Breast, roasted	90	L	L
Breast, untrimmed, roasted	90	L	L
Chump chops, fried	70	L	L
Chump chops, untrimmed, fried	70	L	L
Chump steaks, untrimmed, fried	90	L	L
Kebabs, grilled	90	L	L
Leg, roasted medium	90	L	L
Leg, untrimmed, roasted medium	90	L	L
Leg, roasted well done	90	L	L
Leg, untrimmed, roasted well done	90	L	L
Leg chops, untrimmed, grilled	70	L	L
Leg half fillet, braised	70	L	L
Leg half fillet, untrimmed, braised	70	L	L
Leg half knuckle, pot roasted	90	L	L

Unless otherwise stated, all meat is lean and trimmed, and all chops and cutlets are boned. Steaks are medium-sized. Unless otherwise stated, all dishes are homemade.

ENERGY kcal	ENERGY kJ	FAT g	SATURATED FAT g	PROTEIN g	CARBO-HYDRATE g	FIBRE g
354	1474	21	9.3	41	0	0
276	1157	11	5	44	0	0
359	1494	21	9.6	42	0	0
374	1565	16	7.3	56	0	0
427	1781	24	10.8	53	0	0
259	1088	9	3.2	45	0	0
284	1193	13	5.2	41	0	0
158	662	5	1.9	29	0	0
200	837	10	4.3	27	0	0
182	764	6	2.3	33	0	0
220	918	11	4.7	30	0	0
118	493	7	3.3	14	0	0
171	710	14	6.6	12	0	0
246	1024	17	7.7	24	0	0
323	1338	27	12.9	20	0	0
149	624	8	3.5	20	0	0
216	895	16	7.6	17	0	0
255	1058	18	8.5	22	0	0
259	1079	17	8.1	26	0	0
183	768	8	3.4	27	0	0
216	904	13	5.3	25	0	0
187	786	8	3.3	28	0	0
218	909	12	5.1	27	0	0
155	648	8	3.5	20	0	0
143	597	7	3.2	19	0	0
179	748	12	5.4	18	0	0
181	760	8	3.5	26	0	0

LAMB	AVERAGE PORTION g	GI	GL
Leg half knuckle, untrimmed, pot roasted	90	L	L
Leg joint, roasted	90	L	L
Leg joint, untrimmed, roasted	90	L	L
Leg steaks, grilled	90	L	L
Leg steaks, untrimmed, grilled	90	L	L
Loin chops, grilled	70	L	L
Loin chops, untrimmed, grilled	70	L	L
Loin chops, roasted	70	L	L
Loin chops, untrimmed, roasted	70	L	L
Loin joint, roasted	90	L	L
Loin joint, untrimmed, roasted	90	L	L
Mince, stewed	90	L	L
Neck fillet, grilled	90	L	L
Neck fillet, untrimmed, grilled	90	L	L
Neck fillet strips, stir-fried	90	L	L
Neck fillet strips, untrimmed, stir-fried	90	L	L
Rack of lamb, roasted	90	L	L
Rack of lamb, untrimmed, roasted	90	L	L
Shoulder half bladeside, pot roasted	90	L	L
Shoulder half bladeside, untrimmed, pot roasted	90	L	L
Shoulder half knuckle, braised	90	L	L
Shoulder half knuckle, untrimmed, braised	90	L	L
Shoulder joint, roasted	90	L	L
Shoulder joint, untrimmed, roasted	90	L	L
Shoulder, roasted	90	L	L
Shoulder, untrimmed, roasted	90	L	L
Stewing lamb, stewed	260	L	L
Stewing lamb, untrimmed, stewed	260	L	L

Unless otherwise stated, all meat is lean and trimmed, and all chops and cutlets are boned. Steaks are medium-sized. Unless otherwise stated, all dishes are homemade.

ENERGY kcal	ENERGY kJ	FAT g	SATURATED FAT g	PROTEIN g	CARBO-HYDRATE g	FIBRE g
213	889	12	5.4	25	0	0
189	791	9	3.1	28	0	0
212	887	12	4.2	27	0	0
178	746	8	3.2	26	0	0
208	868	12	5	25	0	0
149	624	7	3.4	20	0	0
214	888	15	7.4	19	0	0
180	754	9	4.3	24	0	0
251	1043	19	9	20	0	0
188	788	10	4.4	25	0	0
273	1133	20	9.5	23	0	0
187	783	11	5.2	22	0	0
256	1063	17	7.8	25	0	0
272	1131	20	9.3	24	0	0
250	1040	18	7.4	22	0	0
271	1124	21	8.9	21	0	0
203	848	12	5.6	24	0	0
327	1355	27	13.3	21	0	0
211	878	13	5.8	24	0	0
292	1212	23	10.8	21	0	0
191	797	11	4.9	23	0	0
272	1130	21	9.7	21	0	0
212	884	12	5.6	26	0	0
254	1056	18	8.4	23	0	0
196	819	11	5	24	0	0
268	1114	20	9.4	22	0	0
624	2600	38	16.9	69	0	0
725	3013	52	23.9	63	0	0

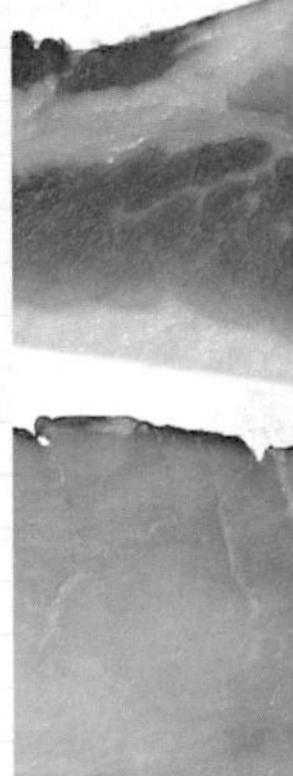

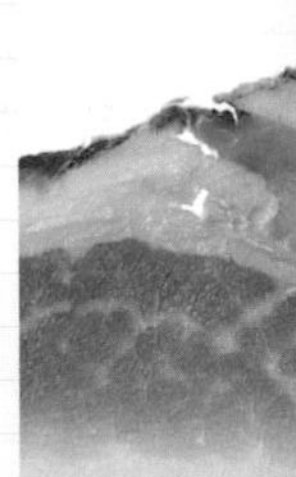

MEAT DISHES

MEAT DISHES	AVERAGE PORTION g	GI	GL
Beef bourguignon	260	M	M
Beef casserole, canned	270	H	H
Beef casserole, made with canned cook-in sauce	300	M	M
Beef hotpot with potatoes, shop bought	260	H	H
Beef stew	280	M	M
Beef stew, made with lean beef	260	M	M
Beef stew and dumplings	260	H	H
Cannelloni, shop bought	260	H	H
Chilli con carne	220	M	M
Chilli con carne, shop bought	220	M	M
Corned beef hash	300	H	H
Cottage pie	310	H	H
Cottage pie, shop bought	310	H	H
Irish stew	330	H	H
Irish stew, made with lean beef	330	H	H
Lamb hotpot with potatoes, shop bought	260	H	H
Lancashire hotpot	260	H	H
Lasagne, shop bought	420	H	H
Meatloaf	100	M	M
Moussaka	330	H	H
Moussaka, shop bought	330	H	H
Pork and beef meatballs in tomato sauce	160	M	M
Pork casserole, made with canned cook-in sauce	260	M	M
Sausage casserole	350	M	M
Shepherd's pie	310	H	H
Shepherd's pie, shop bought	310	H	H
Toad in the hole	160	H	H
Toad in the hole, made with skimmed milk and reduced-fat sausages	160	H	H

Unless otherwise stated, all meat is lean and trimmed, and all chops and cutlets are boned. Steaks are medium-sized. Unless otherwise stated, all dishes are homemade.

ENERGY kcal	ENERGY kJ	FAT g	SATURATED FAT g	PROTEIN g	CARBO-HYDRATE g	FIBRE g
328	1368	17	5.5	36	7	1
211	886	7	3.8	19	19	2.7
408	1707	20	8.1	45	14	2.7
281	1183	11	4.4	19	28	2.3
316	1328	14	4.2	34	14	2
263	1102	9	2.3	32	13	1.8
499	2090	27	12.5	26	41	2.6
315	1326	13	5.2	17	35	3.1
286	1199	17	6.6	20	10	2.4
211	889	9	4.2	17	16	3.1
423	1776	18	9.9	31	37	3
391	1634	21	7.8	20	32	2.8
344	1448	17	7.4	14	37	2.8
399	1676	21	9.6	25	29	3.3
366	1541	17	7.3	26	29	3.3
281	1183	11	4.4	19	28	2.3
289	1206	14	5.5	23	20	2.3
601	2533	26	11.8	31	66	2.9
214	894	11	4.1	17	13	0.5
403	1680	26	11.9	28	15	3.3
462	1934	27	9.6	27	28	2.6
202	840	12	4.4	16	8	1.2
398	1664	20	6.8	44	10	0
578	2405	38	12.3	42	18	3.1
391	1634	21	7.8	20	32	2.8
344	1448	17	7.4	14	37	2.8
443	1853	28	10.7	19	31	1.8
320	1344	14	4.6	20	31	1.9

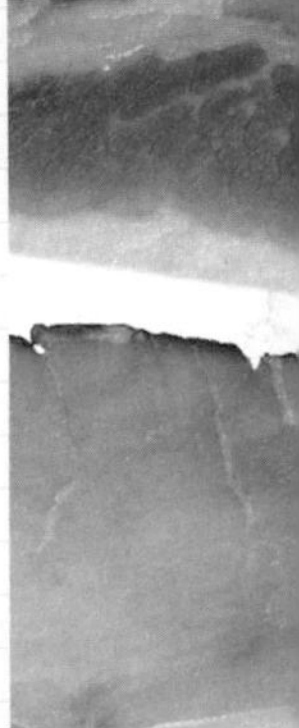

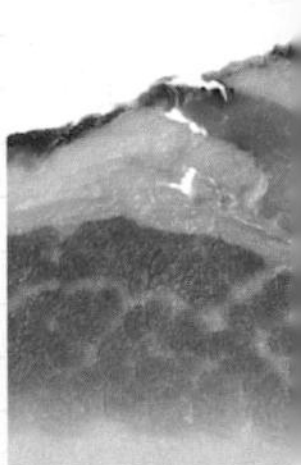

MEAT PRODUCTS	AVERAGE PORTION g	GI	GL
Black pudding, dry-fried	75	M	L
Corned beef	50	M	L
Faggots in gravy, shop bought	150	M	L
Haggis, boiled	85	M	L
Luncheon meat, canned	40	M	L
Meatloaf, shop bought	100	M	L
Scotch eggs, shop bought	120	M	L
White pudding	75	M	L

SAUSAGES			
Beef sausages, fried	40	M	L
Beef sausages, grilled	40	M	L
Bratwurst	75	M	L
Chorizo	30	M	L
Frankfurter	47	M	L
Garlic sausage	40	M	L
Kabana	30	M	L
Mortadella	24	M	L
Pastrami	30	L	L
Pork and beef sausages, grilled	40	M	L
Pork and beef economy sausages, fried	40	M	L
Pork and beef economy sausages, grilled	40	M	L
Pork sausages, fried	40	M	L
Pork sausages, grilled	40	M	L
Pork sausages, frozen, fried	40	M	L
Pork sausages, frozen, grilled	40	M	L
Pork sausages, reduced-fat, fried	40	M	L
Pork sausages, reduced-fat, grilled	40	M	L
Premium sausages, fried	40	M	L
Premium sausages, grilled	40	M	L
Salami	30	L	L
Saveloy, unbattered, shop bought	65	M	L

Unless otherwise stated, all meat is lean and trimmed, and all chops and cutlets are boned. Steaks are medium-sized. Unless otherwise stated, all dishes are homemade.

ENERGY kcal	ENERGY kJ	FAT g	SATURATED FAT g	PROTEIN g	CARBO-HYDRATE g	FIBRE g
223	927	16	6.4	8	12	0.2
103	430	5	2.8	13	1	0
222	929	11	3.8	12	19	0.3
264	1098	18	6.5	9	16	0.2
112	463	10	3.5	5	1	0.1
216	900	16	5.6	17	1	0.5
301	1255	5.2	19	14	16	0
338	1407	24	8	5	27	1
112	464	8	3	5	5	0.3
111	463	8	3.2	5	5	0.3
195	811	16	6	12	2	Trace
87	362	7	2.9	5	1	Trace
135	559	12	4.3	6	1	0
99	412	8	2.9	6	0	Trace
92	383	8	3.4	5	0	Trace
79	328	7	2.8	3	0	Trace
37	155	1.3	0.5	5.8	0.5	0
108	448	8	3	5	4	0.5
99	414	7	1.8	5	5	0.5
103	429	7	1.9	5	5	0.5
123	512	10	3.4	6	4	0.3
118	488	9	3.2	6	4	0.3
126	525	10	3.5	6	4	0.3
116	482	8	3	6	4	0.3
84	352	5	1.7	6	4	0.6
92	384	6	2	6	4	0.6
110	457	8	3.1	6	3	0.3
117	486	9	3.3	7	3	0.3
131	544	11.8	4.4	6.3	0.1	0
192	801	18	3.6	9	7	0.5

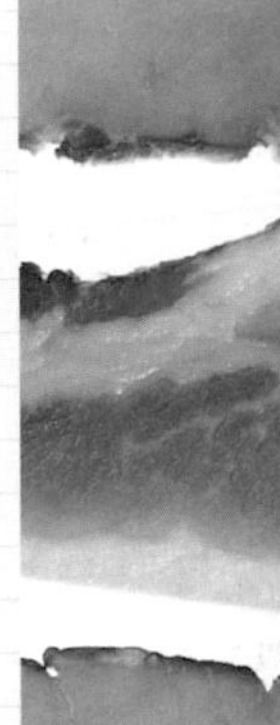

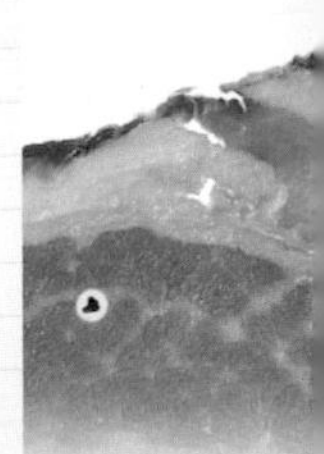

BEEFBURGERS	AVERAGE PORTION g	GI	GL
Beefburgers, fried	100	M	L
Beefburgers, grilled	100	M	L
Beefburgers, low-fat, fried	100	M	L
Beefburgers, low-fat, grilled	100	M	L
Beefburgers in gravy, canned	100	M	L

PIES AND PASTIES			
Beef pie with pastry, individual	150	H	H
Cornish pasty	155	H	H
Cornish pasty, shop bought	155	H	H
Lamb samosa, baked	70	H	H
Lamb samosa, deep-fried	70	H	H
Pork and egg pie	60	H	H
Pork pie	60	H	H
Pork pie, mini	50	H	H
Sausage rolls, flaky pastry	60	H	H
Sausage rolls, shortcrust pastry	60	H	H
Steak and kidney pie, double crust	120	H	H
Steak and kidney pie, shop bought	141	H	H

OFFAL			
Heart, lamb, roasted	200	L	L
Heart, lamb, stewed	200	L	L
Heart, ox, stewed	200	L	L
Heart, pig, stewed	100	L	L
Kidney, lamb, fried	35	L	L
Kidney, ox, stewed	112	L	L
Kidney, pig, fried	140	L	L
Kidney, pig, stewed	140	L	L
Liver, calf, fried	100	L	L
Liver, chicken, fried	70	L	L
Liver, lamb, fried	100	L	L

Unless otherwise stated, all meat is lean and trimmed, and all chops and cutlets are boned. Steaks are medium-sized. Unless otherwise stated, all dishes are homemade.

ENERGY kcal	ENERGY kJ	FAT g	SATURATED FAT g	PROTEIN g	CARBO-HYDRATE g	FIBRE g
303	1261	23	8.7	24	1	0.2
287	1194	20	8.8	25	1	0.2
193	807	11	5	24	0	0
178	745	10	4.4	23	1	0
171	713	12	4.8	12	5	Trace
438	1830	26.6	11.9	13.8	38.3	2
431	1800	27.6	13.2	10.9	37.2	1.7
414	1731	25	9.1	10	39	1.4
187	781	10	2.5	8	16	1.2
265	1097	22	3.3	6	12	0.8
178	740	13	4.4	6	10	0.5
222	925	15.6	6.1	5.9	15.4	0.7
196	815	14	5.7	5	13	0.5
238	991	17	6.5	6	15	0.8
229	953	16	5.8	6	17	0.9
406	1694	26	9.7	16	27	1.1
437	1826	24	11.8	12	38	0.7
452	1888	28	6.2	51	0	0
312	1308	15	6	24	20	0
314	1322	10	5	56	0	0
162	678	7	1.3	25	0	0
66	274	4	0.4	8	0	0
155	648	5	1.6	27	0	0
283	1187	13	2.1	41	0	0
214	897	9	2.8	34	0	0
176	734	10	2.7	22	Trace	0
118	494	6	3.5	15	Trace	0
237	989	13	4.9	30	Trace	0

OFFAL	AVERAGE PORTION g	GI	GL
Liver, ox, stewed	70	L	L
Liver, pig, stewed	70	L	L
Liver and bacon, fried	100	L	L
Liver and onions, stewed	142	L	L
Liver sausage	40	L	L
Oxtail, stewed	260	L	L
Sweetbread, lamb, fried	100	L	L
Tongue	50	L	L
Tongue, canned	50	L	L
Tongue, ox, stewed	50	L	L
Tongue, sheep, stewed	50	L	L
Tripe and onions, stewed	100	L	L

PORK, BACON AND HAM			
Bacon			
Collar joint, boiled	46	L	L
Collar joint, untrimmed, boiled	46	L	L
Loin steaks, grilled	100	L	L
Rashers, back, dry-cured, grilled	100	L	L
Rashers, back, dry-fried	100	L	L
Rashers, back, grilled	100	L	L
Rashers, back, untrimmed, grilled	100	L	L
Rashers, back, grilled crispy	100	L	L
Rashers, back, reduced salt, grilled	100	L	L
Rashers, back, smoked, grilled	100	L	L
Rashers, back, sweetcure, grilled	100	L	L
Rashers, back, 'tendersweet', grilled	100	L	L
Rashers, middle, fried	100	L	L
Rashers, middle, grilled	100	L	L
Rashers, streaky, fried	100	L	L
Rashers, streaky, grilled	100	L	L
Belly joint, untrimmed, roasted	110	L	L
Belly, untrimmed, grilled	110	L	L

Unless otherwise stated, all meat is lean and trimmed, and all chops and cutlets are boned. Steaks are medium-sized. Unless otherwise stated, all dishes are homemade.

ENERGY kcal	ENERGY kJ	FAT g	SATURATED FAT g	PROTEIN g	CARBO-HYDRATE g	FIBRE g
139	582	7	2.5	17	2	0
132	555	6	1.8	18	2	0
247	1030	15	2.1	29	Trace	0
210	879	11	2.1	21	8	1.4
90	377	7	2.1	5	2	0.3
632	2636	35	12	79	0	0
217	910	11	5	29	0	0
101	418	7	3	9	Trace	Trace
107	442	8	3.2	8	Trace	0
121	504	9	3	10	0	0
145	599	12	5	9	0	0
93	393	3	1.5	8	10	0.7
88	368	4	1.7	12	0	0
150	619	12	4.9	9	0	0
191	799	10	3.5	26	0	0
257	1071	16	6	28	0	0
295	1225	22	8.3	24	0	0
214	892	12	4.6	26	0	0
287	1194	22	8.1	23	0	0
313	1308	19	7.1	36	0	0
282	1172	21	7.8	24	0	0
293	1216	22	8.3	23	0	0
258	1074	17	6.6	24	2	0
213	889	12	4.5	26	Trace	0
350	1452	29	9.8	23	0	0
307	1276	23	8.4	25	0	0
335	1389	27	9.1	24	0	0
337	1400	27	9.8	24	0	0
322	1341	24	8.1	28	0	0
352	1465	26	9	30	0	0

PORK, BACON AND HAM	AVERAGE PORTION g	GI	GL
Chump chops, untrimmed, fried	170	L	L
Chump steaks, untrimmed, fried	170	L	L
Crackling, cooked	50	L	L
Fillet, grilled, lean	120	L	L
Fillet, untrimmed, grilled	120	L	L
Fillet strips, stir-fried, lean	90	L	L
Gammon			
Joint, boiled	170	L	L
Rashers, grilled	100	L	L
Ham			
Ham, canned	90	L	L
Ham, Pancetta	40	L	L
Ham, Parma	30	L	L
Ham, premium	56	L	L
Ham, Prosciutto	30	L	L
Ham, Serrano	30	L	L
Pork shoulder, cured	100	L	L
Kebabs, grilled	90	L	L
Kebabs, untrimmed, grilled	90	L	L
Leg joint, untrimmed, frozen, roasted	90	L	L
Leg joint, roasted medium	90	L	L
Leg joint, untrimmed, roasted medium	90	L	L
Leg joint, roasted well done	90	L	L
Leg joint, untrimmed, roasted well done	90	L	L
Loin chops, grilled	75	L	L
Loin chops, untrimmed, grilled	75	L	L
Loin chops, grilled	120	L	L
Loin chops, untrimmed, grilled	120	L	L
Loin chops, roasted	75	L	L
Loin chops, untrimmed, roasted	75	L	L
Loin joint, pot roasted	90	L	L
Loin joint, untrimmed, pot roasted	90	L	L
Loin joint, roasted	90	L	L
Loin joint, untrimmed, roasted	90	L	L

Unless otherwise stated, all meat is lean and trimmed, and all chops and cutlets are boned. Steaks are medium-sized. Unless otherwise stated, all dishes are homemade.

ENERGY kcal	ENERGY kJ	FAT g	SATURATED FAT g	PROTEIN g	CARBO-HYDRATE g	FIBRE g
498	2069	37	11.9	42	0	0
369	1542	20	6.3	47	0	0
275	1140	23	7.7	18	0	0
204	863	5	1.8	40	0	0
214	899	6	2.3	40	0	0
164	688	5	1.2	29	0	0
347	1447	21	7	40	0	0
199	834	10	3.4	28	0	0
96	404	4	1.4	15	0	0
137	572	11.9	1.8	7.2	0.2	0
67	280	3.8	1.3	8.2	3.8	0
74	310	3	1	12	0	0
56	234	2.5	0.8	8.4	0.1	0
96	402	9	1.5	3.8	0.2	0
103	435	4	1.2	17	1	0
161	679	4	1.4	31	0	0
170	717	5	2	30	0	0
202	842	10	3.6	28	0	0
164	689	5	1.7	30	0	0
194	813	9	3.2	28	0	0
167	701	5	1.6	31	0	0
233	974	14	4.9	27	0	0
140	585	5	1.8	23	0	0
191	800	12	4.3	21	0	0
221	929	8	2.6	38	0	0
308	1289	19	6.7	35	0	0
181	758	8	2.8	28	0	0
226	942	14	5.3	24	0	0
177	744	7	2.6	28	0	0
275	1142	21	7.4	22	0	0
164	687	6	2.2	27	0	0
228	949	15	5.3	24	0	0

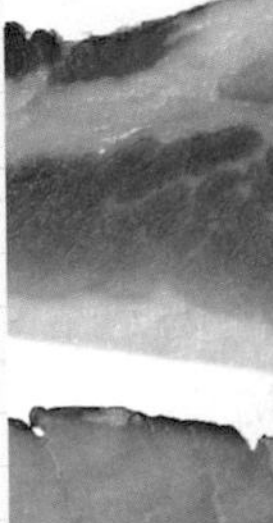

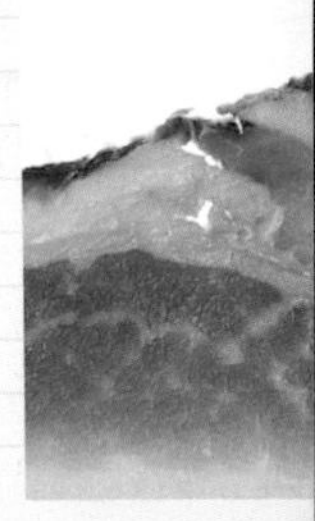

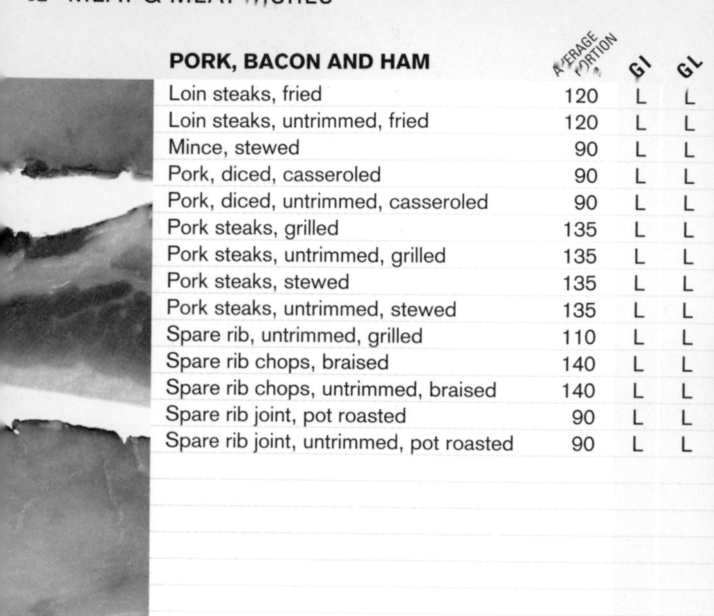

PORK, BACON AND HAM	AVERAGE PORTION	GI	GL
Loin steaks, fried	120	L	L
Loin steaks, untrimmed, fried	120	L	L
Mince, stewed	90	L	L
Pork, diced, casseroled	90	L	L
Pork, diced, untrimmed, casseroled	90	L	L
Pork steaks, grilled	135	L	L
Pork steaks, untrimmed, grilled	135	L	L
Pork steaks, stewed	135	L	L
Pork steaks, untrimmed, stewed	135	L	L
Spare rib, untrimmed, grilled	110	L	L
Spare rib chops, braised	140	L	L
Spare rib chops, untrimmed, braised	140	L	L
Spare rib joint, pot roasted	90	L	L
Spare rib joint, untrimmed, pot roasted	90	L	L

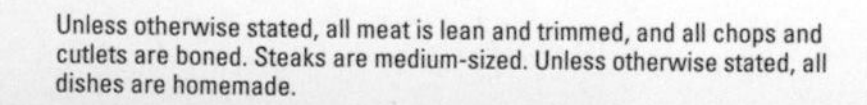

Unless otherwise stated, all meat is lean and trimmed, and all chops and cutlets are boned. Steaks are medium-sized. Unless otherwise stated, all dishes are homemade.

ENERGY kcal	ENERGY kJ	FAT g	SATURATED FAT g	PROTEIN g	CARBO-HYDRATE g	FIBRE g
229	962	9	2.8	38	0	0
331	1378	22	7.2	33	0	0
172	720	9	3.5	22	0	0
166	698	6	1.7	29	0	0
166	698	6	1.8	28	0	0
228	963	5	1.8	46	0	0
267	1123	10	3.6	44	0	0
238	1000	6	1.8	45	0	0
269	1126	11	3.2	42	0	0
321	1340	21	8.3	32	0	0
298	1249	14	5	43	0	0
346	1446	21	7.7	39	0	0
181	759	8	2.9	27	0	0
234	973	16	6	22	0	0

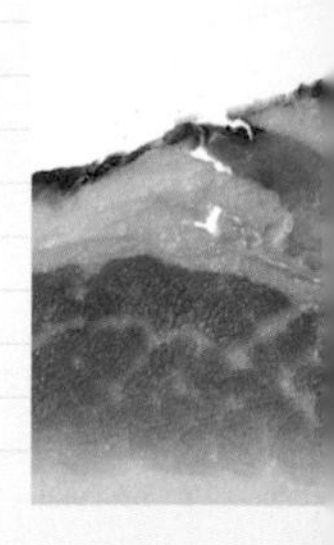

CHICKEN	AVERAGE PORTION g	GI	GL
Breast, casseroled	130	L	L
Breast, skinned, casseroled	130	L	L
Breast, grilled	130	L	L
Breast, grilled and skinned	130	L	L
Breast, skinned, grilled	130	L	L
Breast strips, stir-fried	90	L	L
Corn-fed chicken, roasted, dark meat	90	L	L
Corn-fed chicken, roasted, light meat	90	L	L
Dark meat, roasted	100	L	L
Drumsticks, casseroled	47	L	L
Drumsticks, skinned, casseroled	47	L	L
Drumsticks, roasted	47	L	L
Drumsticks, skinned, roasted	47	L	L
Leg quarter, casseroled	146	L	L
Leg quarter, skinned, casseroled	146	L	L
Leg quarter, roasted	146	L	L
Light meat, roasted	100	L	L
Thighs, casseroled	45	L	L
Thighs, skinned, boned and diced, casseroled	45	L	L
Wing quarter, casseroled	150	L	L
Wing quarter, skinned, casseroled	150	L	L
Wing quarter, roasted	150	L	L
Wings, grilled	100	L	L

CHICKEN PRODUCTS			
Breaded/battered chicken pieces,	100	H	H
Breaded chicken strips	100	H	H
Chicken fingers, baked	90	H	H
Chicken goujons, baked	90	H	H
Chicken in breadcrumbs, stuffed with cheese and vegetables, baked	100	H	H
Chicken in white sauce, canned	100	L	L

Unless otherwise stated, chicken and turkey are neither skinned nor boned, game is skinned and trimmed and dishes are homemade.

ENERGY kcal	ENERGY kJ	FAT g	SATURATED FAT g	PROTEIN g	CARBO-HYDRATE g	FIBRE g
239	1004	11	3.1	35	0	0
148	628	7	2	37	0	0
225	946	8	1.2	38	0	0
191	807	4	1.2	39	0	0
192	814	4	0.8	42	0	0
145	609	4	0.8	27	0	0
167	695	9	2.5	22	0	0
127	536	4	1.1	23	0	0
196	819	11	2.9	24	0	0
102	425	7	1.8	10	0	0
87	363	5	1.2	11	0	0
87	364	4	1.2	12	0	0
71	301	2	0.7	13	0	0
317	1320	20	5.5	33	0	0
257	1075	12	3.4	37	0	0
345	1432	25	6.7	31	0	0
153	645	4	1	30	0	0
105	436	7	2	10	0	0
81	340	4	1.1	12	0	0
315	1316	19	5.3	37	0	0
246	1035	9	2.6	40	0	0
339	1415	21	5.9	37	0	0
274	1146	17	4.6	27	Trace	0
256	1073	13.9	2.1	14.4	19.6	1.1
277	1161	14	4	19.4	19.6	0.7
185	774	9	2.7	11	17	Trace
249	1045	13	3.6	17	18	0.6
230	963	14	4.1	16	11	0.9
141	590	8	2.3	14	3	Trace

CHICKEN PRODUCTS

	AVERAGE PORTION g	GI	GL
Chicken Kiev, frozen, baked	170	H	H
Chicken pancakes, frozen, shallow-fried	100	M	H
Chicken roll	24	L	L
Chicken slices	80	L	L
Coated chicken burger, baked	100	H	H
Coated chicken steak, baked	100	H	H

CHICKEN PIES

	AVERAGE PORTION g	GI	GL
Chicken and mushroom pie, single crust	100	H	H
Chicken pasty	155	H	H
Chicken pie, individual, baked	130	H	H

CHICKEN DISHES

	AVERAGE PORTION g	GI	GL
Chicken chasseur	260	L	L
Chicken in sauce with vegetables	290	L	L
Chicken curry	350	L	L
Chicken curry, with bone	350	L	L
Chicken curry, without bone	300	L	L
Chicken fricassée	200	L	M
Chicken in white sauce, made with semi-skimmed milk	200	L	M
Chicken in white sauce, made with whole milk	200	L	L
Chicken risotto	350	M	H
Coronation chicken	200	L	L
Lemon chicken	100	L	L
Tandoori chicken	100	L	L

Unless otherwise stated, chicken and turkey are neither skinned nor boned, game is skinned and trimmed and dishes are homemade.

ENERGY kcal	ENERGY kJ	FAT g	SATURATED FAT g	PROTEIN g	CARBO-HYDRATE g	FIBRE g
456	1902	29	12.1	32	19	1
260	1090	14	1.4	6	29	0.1
31	132	1	0.4	4	1	Trace
91	386	1	0.3	19	2	0
266	1113	15.5	2.6	14.2	18.7	1.1
234	982	11.6	1.8	17.7	15.8	1.1
200	836	10	4.5	13	14	0.6
448	1868	28.7	14.2	12.6	37	1.6
374	1563	21	9.1	12	32	1
203	861	5	0.8	33	7	0.8
336	1412	15	7	39	13	0.9
522	2174	31	14	42	19	4.5
539	2237	44	6	27	8	2.4
615	2550	51	6.6	31	9	2.7
214	896	12	4.8	22	6	1
310	1298	16	5	34	10	0.2
328	1376	18	6.2	34	10	0.2
546	2310	10	4.5	31	84	Trace
728	3012	63	10.4	33	6	Trace
155	652	6	0.8	16	9	Trace
214	897	11	3.3	27	2	Trace

GAME	AVERAGE PORTION g	GI	GL
Duck, roasted	185	L	L
Duck, untrimmed, roasted	185	L	L
Goose, roasted	185	L	L
Goose, untrimmed, roasted	185	L	L
Grouse, roasted	160	L	L
Pheasant, roasted	160	L	L
Pigeon, roasted	115	L	L
Rabbit, stewed	160	L	L
Venison, roasted	120	L	L

TURKEY			
Breast, grilled and skinned	90	L	L
Dark meat, roasted	90	L	L
Drumsticks, roasted	90	L	L
Drumsticks, roasted and skinned	90	L	L
Light meat, from self-basting bird, roasted	90	L	L
Light meat, roasted	90	L	L
Mince, stewed	90	L	L
Strips, stir-fried	90	L	L
Thighs, diced, skinned, boned, casseroled	90	L	L

Unless otherwise stated, chicken and turkey are neither skinned nor boned, game is skinned and trimmed and dishes are homemade.

ENERGY kcal	ENERGY kJ	FAT g	SATURATED FAT g	PROTEIN g	CARBO-HYDRATE g	FIBRE g
361	1508	19	6.1	47	0	0
783	3238	92	21.1	37	0	0
590	2455	41	13.7	54	0	0
557	2316	39	12.2	51	0	0
205	869	3	0.8	44	0	0
352	1469	5	0.6	45	0	0
215	903	9	3	33	0	0
182	766	5	2.7	34	0	0
198	838	3	1	43	0	0
140	592	2	0.5	32	0	0
159	671	6	1.8	26	0	0
167	702	8	2.3	25	0	0
146	615	5	1.5	25	0	0
147	619	4	1.1	29	0	0
138	583	2	0.6	30	0	0
158	665	6	1.8	26	0	0
148	623	4	1.1	28	0	0
163	684	7	2.3	25	0	0

FISH	AVERAGE PORTION g	GI	GL
Anchovies, canned in oil, drained	3	L	L
Cod, baked	120	L	L
Cod, coated in batter, frozen, baked	180	H	M
Cod, coated in batter, takeaway	180	H	H
Cod, coated in breadcrumbs, baked	180	H	H
Cod, dried, salted, boiled	90	L	L
Cod, frozen, grilled	120	L	L
Cod, in parsley sauce, frozen, boiled	170	M	M
Cod, microwaved	120	L	L
Cod, poached	120	L	L
Cod, smoked, poached	120	L	L
Cod, steamed	120	L	L
Conger eel, grilled	115	L	L
Dogfish, coated in batter, fried	125	H	H
Eels, jellied	70	L	L
Eels, stewed	70	L	L
Haddock, coated in batter, fried	120	H	H
Haddock, coated in breadcrumbs, fried	120	H	H
Haddock, coated in flour, fried	120	H	H
Haddock, grilled	120	L	L
Haddock, poached	120	L	L
Haddock, smoked, poached	150	L	L
Haddock, smoked, steamed	150	L	L
Haddock, steamed	120	L	L
Hake, grilled	100	L	L
Halibut, grilled	145	L	L
Halibut, poached	110	L	L
Halibut, steamed	110	L	L
Herring, grilled	119	L	L
Herring, coated in oatmeal, fried	119	M	M
Herring, pickled	90	L	L
Kipper, baked	130	L	L
Kippers, boiled in bag with butter	170	L	L

Unless otherwise stated, values for bottled and canned seafood are for drained weights. Dishes are homemade unless otherwise stated.

ENERGY kcal	ENERGY kJ	FAT g	SATURATED FAT g	PROTEIN g	CARBO-HYDRATE g	FIBRE g
6	24	0.3	0	0.8	0	0
120	510	0.6	0.1	28.7	0	0
380	1589	21	6.5	23	26	1.1
432	1802	26.5	13.6	30.2	19.3	0.9
367	1544	14.9	2.4	24.7	35.6	3.1
124	527	1	0.2	29	0	0
114	482	2	0.5	25	Trace	0
143	598	5	3	20	5	0.2
118	497	0.5	0.1	28.2	0	0
113	475	1	0.4	25	Trace	0
121	511	2	0.7	26	Trace	0
100	420	1	0.2	22	0	0
158	660	6	Trace	25	0	0
369	1531	27	6.6	18	13	0.5
69	284	5	1.3	6	Trace	0
141	589	9	1.3	14	0	0
278	1163	17	4.4	21	12	0.5
209	875	10	0.8	26	4	0.2
166	698	5	0.5	25	5	0.2
118	500	0.4	0.1	28.7	0	0
136	572	5	3.1	21	1	0
138	584	0.8	0.1	32.7	0	0
152	644	1	0.3	35	0	0
112	472	0.7	0.1	26.2	0	0
113	478	3	0.4	22	0	0
175	744	3	0.6	37	0	0
169	713	6	3	27	1	0
144	608	4	0.6	26	0	0
215	900	13	3.3	24	0	0
278	1160	18	3.3	27	2	0.1
188	789	10	3.3	15	9	0
267	1112	15	2.3	33	0	0
328	1369	22.4	5	31.6	0	0

FISH	AVERAGE PORTION g	GI	GL
Kippers, grilled	130	L	L
Lemon sole, goujons, baked	170	H	H
Lemon sole, goujons, fried	170	H	H
Lemon sole, grilled	109	L	L
Lemon sole, steamed	109	L	L
Mackerel, canned in brine	45	L	L
Mackerel, canned in tomato sauce	125	L	L
Mackerel, fried	161	L	L
Mackerel, grilled	140	L	L
Mackerel, smoked	150	L	L
Monkfish, grilled	70	L	L
Mullet, grey, grilled	100	L	L
Mullet, red, grilled	110	L	L
Pilchards, canned in tomato sauce	215	L	L
Plaice, coated in batter, fried	200	H	M
Plaice, coated in breadcrumbs, baked	150	H	H
Plaice, coated in breadcrumbs, fried	150	H	H
Plaice, frozen, grilled	165	L	L
Plaice, frozen, steamed	130	L	L
Plaice, goujons, baked	130	H	H
Plaice, goujons, fried	150	H	H
Plaice, grilled	130	L	L
Plaice, steamed	130	L	L
Red snapper, fried	102	L	L
Rock salmon, in batter, fried	125	H	H
Salmon, cold smoked	56	L	L
Salmon, grilled/baked	100	L	L
Salmon, hot smoked	56	L	L
Salmon, pink, canned	45	L	L
Salmon, raw	100	L	L
Salmon, red, canned	45	L	L
Salmon, steamed	77	L	L
Sardines, canned in brine	45	L	L

Unless otherwise stated, values for bottled and canned seafood are for drained weights. Dishes are homemade unless otherwise stated.

ENERGY kcal	ENERGY kJ	FAT g	SATURATED FAT g	PROTEIN g	CARBO-HYDRATE g	FIBRE g
319	1326	22.9	4.9	28.2	0	0
318	1318	25	Trace	27	25	Trace
636	2640	49	5.4	26	24	Trace
106	445	2	0.2	22	0	0
99	419	1	0.1	22	0	0
92	382	6.3	1.4	8.6	0	0
258	1070	19	4.1	20	2	Trace
438	1819	31	6.4	39	0	0
396	1644	31.4	7.1	28.4	0	0
452	1892	36.2	7.6	31.7	0	0
67	285	0	0.1	16	0	0
150	629	5	1.4	26	0	0
133	561	5	1.4	22	0	0
310	1292	17	3.7	36	2	Trace
514	2144	34	9	30	24	1
365	1527	17.4	1.8	21.6	32.3	0.8
342	1427	21	2.3	27	13	0.3
200	843	3	0.5	42	0	0
120	506	2	0.3	25	0	0
395	1651	24	0	11	36	Trace
639	2657	48	5.4	13	41	Trace
125	525	2	0.4	26	0	0
121	510	2	0.4	25	0	0
129	542	3	0.7	25	0	0
369	1531	27	6.6	18	13	0.5
103	431	5.7	1.2	12.8	0.3	0
166	694	7.6	1.3	24	0.5	0
104	435	4.9	1.1	14.2	0.7	0
62	261	2.2	0.4	10.6	0	0
180	750	11	1.9	20	0	0
72	301	3.3	0.6	10.6	0	0
152	634	10	1.8	15	0	0
77	320	4.1	1.2	4.1	0	0

FISH	AVERAGE PORTION g	GI	GL
Sardines, canned in oil	100	L	L
Sardines, canned in tomato sauce	45	M	L
Sardines, grilled	40	L	L
Sea bass, baked	100	L	L
Skate, coated in batter, fried	170	H	M
Skate, grilled	215	L	L
Sole, grilled	120	L	L
Sprats, fried	132	L	L
Swordfish, grilled	125	L	L
Trout, brown, steamed	155	L	L
Trout, rainbow, baked	155	L	L
Trout, rainbow, grilled	155	L	L
Tuna, canned in brine	45	L	L
Tuna, canned in sunflower oil	45	L	L
Tuna, grilled/baked	100	L	L
Tuna, raw	45	L	L
Whitebait, coated in flour, fried	80	H	H
Whiting, coated in breadcrumbs, fried	162	H	H
Whiting, steamed	85	L	L

FISH PRODUCTS AND DISHES			
Fish balls, steamed	50	L	L
Fish cakes, fried	100	H	H
Fish cakes, breaded, baked	100	H	H
Fish cakes, salmon, breaded, baked	100	H	H
Fish fingers, cod, fried	56	H	H
Fish fingers, cod, grilled/baked	100	H	H
Fish fingers, pollack, grilled	100	H	H
Fish fingers, salmon, grilled/baked	100	H	H
Fish pie	170	H	H
Kedgeree	300	H	H
Roe, cod, hard, coated in batter, fried	160	L	L
Roe, cod, hard, fried	116	L	L

Unless otherwise stated, values for bottled and canned seafood are for drained weights. Dishes are homemade unless otherwise stated.

ENERGY kcal	ENERGY kJ	FAT g	SATURATED FAT g	PROTEIN g	CARBO-HYDRATE g	FIBRE g
220	918	14	2.9	23	0	0
79	328	4.9	1.3	8.3	0.4	0
78	326	4	1.2	10	0	0
154	646	6.8	1.5	23.2	0	0
286	1193	17	4.3	25	8	0.3
170	725	1	0.2	41	0	0
109	466	0.7	0.2	25.8	0	0
548	2268	46	7	33	0	0
174	729	6	1.5	29	0	0
209	877	7	1.6	36	0	0
233	977	9.5	2.2	36.9	0	0
209	876	8	1.7	33	0	0
49	205	0.5	0.1	11.2	0	0
72	300	2.9	0.4	11.4	0	0
146	611	4	0.8	27.3	0.4	0
48	204	0.3	0.1	11.3	0	0
420	1739	38	2.1	16	4	0.2
309	1298	17	1.8	29	11	0.3
78	331	1	0.1	18	0	0
37	157	0	0.9	6	3	0
218	912	14	1.4	8	16	0.6
206	867	9.4	1	9.3	22.6	0.4
245	1027	13.7	2	11.4	20.4	0.4
133	557	8	2	7	9	0.3
223	932	9.2	1.2	14.3	22	0.7
213	897	9.2	1.2	13.9	20	1.8
247	1035	11.2	1.1	17.2	20.7	0.7
210	880	8.2	4.3	11.2	24.5	1
498	2103	24	6.9	43	32	Trace
302	1264	19	2.6	20	14	0.3
234	979	14	1.9	24	3	0.1

FISH PRODUCTS AND DISHES	AVERAGE PORTION g	GI	GL
Roe, herring, soft, fried	85	L	L
Salmon en croûte, shop bought	100	H	H

SEAFOOD AND SHELLFISH			
Cockles, boiled	25	L	L
Cockles, bottled in vinegar	25	L	L
Crab, boiled, dressed with shell	130	L	L
Crab, brown meat, cooked	50	L	L
Crab, canned in brine	40	L	L
Crab, white meat, cooked	50	L	L
Langoustines, boiled	100	L	L
Lobster, boiled, dressed with shell	250	L	L
Mussels, canned or bottled without shells	40	L	L
Mussels, cooked	100	L	L
Mussels in white wine sauce	100	L	L
Oysters, uncooked and shelled	120	L	L
Prawns, cooked	60	L	L
Prawns, king, cooked	100	L	L
Prawns, king, grilled	100	L	L
Scallops, grilled/baked	100	L	L
Scallops, steamed and shelled	70	L	L
Squid, coated in batter, fried	120	H	H
Squid, grilled/baked	100	L	L
Whelks, boiled and shelled	30	L	L
Winkles, boiled and shelled	30	L	L

SEAFOOD PRODUCTS AND DISHES			
Calamari, coated in batter, baked	100	H	H
Crabsticks	100	M	L
Scampi coated in breadcrumbs, baked	170	H	H
Seafood cocktail	88	L	L
Seafood sticks	100	L	L

Unless otherwise stated, values for bottled and canned seafood are for drained weights. Dishes are homemade unless otherwise stated.

ENERGY kcal	ENERGY kJ	FAT g	SATURATED FAT g	PROTEIN g	CARBO-HYDRATE g	FIBRE g
225	941	13	2.2	22	4	0.2
288	1202	19	3.1	12	18	0.3
13	57	1	0.1	3	Trace	0
15	63	1	0.1	3	Trace	0
166	696	7	0.9	25	Trace	0
73	304	3.9	0.6	9.4	0.5	0
31	130	0	0	7	Trace	0
43	180	0.2	0	10.3	0.5	0
86	369	0.8	0.2	0.8	0	0
258	1088	4	0.5	55	Trace	0
39	166	1	0.2	7	1	0
104	438	2.2	0.3	17.7	0	0
81	341	3.2	1.3	9.7	3.7	0.1
78	330	2	0.2	13	3	0
42	177	0.5	0.1	9.2	0	0
68	290	0.4	0.1	16.2	0	0
102	433	0.9	0.2	23.5	0	0
127	531	3.9	0.6	20.2	2.9	0
83	351	1	0.3	16	2	Trace
234	978	12	2.5	14	19	0.6
131	548	4.6	1	18.8	3.8	0
27	113	1	0.1	6	Trace	0
22	92	1	0.1	5	Trace	0
223	1206	17.5	2.1	8.5	25.9	0.4
68	290	0	0	10	7	0
394	1658	17.9	1.4	19.7	41.3	0.5
77	325	1	0.3	14	3	0
102	428	1.9	0.3	7.3	14.9	0

INDIAN	AVERAGE PORTION g	GI	GL
Bombay aloo	100	H	H
Chicken biryani	400	H	H
Chicken, butter/murgh makhani	350	L	L
Chicken dhansak	350	L	L
Chicken dupiaza	350	L	L
Chicken jalfrezi	350	L	L
Chicken korma	350	L	L
Chicken madras	350	L	L
Chicken tikka	350	L	L
Chicken tikka masala	350	L	L
Chicken vindaloo	350	L	L
Lamb balti	350	L	L
Lamb rogan josh	350	L	L
Meat samosas	70	H	H
Naan bread	50	M	H
Poppadoms	70	M	H
Prawn bhuna	400	L	L
Prawn madras	350	L	L
Raita	50	L	L
Saag paneer	350	L	L
Vegetable balti	350	L	L
Vegetable bhaji	100	H	H
Vegetable biryani	350	H	H

SOUTH-EAST ASIAN DISHES			
Aromatic crispy duck	125	L	L
Barbecue pork bun	100	H	H
Barbecue roast pork	100	L	L
California roll	100	H	H
Chicken chow mein	350	H	H
Chicken fried rice	350	H	H
Chicken satay	170	L	L
Chicken with cashew nuts	360	L	L

ENERGY kcal	ENERGY kJ	FAT g	SATURATED FAT g	PROTEIN g	CARBO-HYDRATE g	FIBRE g
118	492	6.7	0.8	2	13.8	1.2
651	2144	30	8	34	66	4.4
666	2786	50.4	26.3	45.5	18.6	3.5
503	1509	30	6	40	21	6.7
397	1168	25	4.9	40	4	6.3
416	1203	27	5.3	34	11	7
668	1753	51	20	44	8	7
408	1707	22	9.8	44.4	9.4	7.8
421	1470	15	5.6	71	Trace	1
551	1488	40	13.7	47	Trace	7.7
424	1782	14.3	2.6	68.6	9.5	1.1
534	1518	35	8.8	43	11	6.7
510	1422	35	9.8	43	6	3.1
191	553	12	3.2	8	13	1.5
142	602	3.6	0.5	3.9	25.1	1
351	916	27	5.6	8	20	4.1
363	802	35	4	6	7	7.6
390	1042	29	3.5	27	6	7
57	241	2.2	1.5	4.2	5.8	0.7
412	1723	18.8	6	12.1	51.7	7.7
373	983	28	5.3	8	23	7.7
166	688	13.7	1.5	3.4	8.2	1.9
467	1468	25	4.9	10	55	6
412	1107	30	9.1	35	15	1.1
273	1143	9.9	3.2	9.4	39	1.4
207	867	13.4	5.5	18.8	3	0.4
116	485	0.5	0.1	5.3	24.3	0.9
516	1662	25	4.2	30	46	3.9
562	1948	21	3.5	23	75	0
324	1006	18	4.9	37	5	3.7
470	1331	31	5.4	38	10	0

SOUTH-EAST ASIAN DISHES	AVERAGE PORTION g	GI	GL
Chinese egg tart	50	M	H
Chinese roast duck with skin	100	L	L
Egg fried rice	270	H	H
Green chicken curry	350	L	L
Meat spring roll	55	H	H
Pancakes	70	H	H
Prawn crackers	70	M	M
Salmon maki	100	H	H
Salmon nigiri sushi	100	H	H
Sesame prawn toasts	70	M	M
Spare ribs	340	L	L
Steamed barbecue pork bun	50	H	H
Steamed chicken bun	50	H	H
Steamed pork dumpling, 'siu mai'	30	L	L
Steamed prawn dumpling, 'har gow'	30	L	L
Steamed rice roll with barbecue pork	100	H	H
Steamed rice roll with prawns	100	H	H
Steamed sticky rice	100	H	H
Stir-fried beef with green peppers in black bean sauce	360	L	L
Stir-fried chicken with vegetables	200	L	L
Stir-fried Thai vegetable curry	350	L	L
Stir-fried vegetables	340	L	L
Sweet and sour chicken	300	M	M
Sweet and sour pork, battered	300	H	H
Szechuan prawns with vegetables	350	L	L
Thai red chicken curry	350	L	L
Tuna maki	100	H	H
Tuna nigiri sushi	100	H	H
Vegetable maki	100	H	H
Wonton, deep-fried	30	M	M
Wonton noodles in soup	225	H	H

ENERGY kcal	ENERGY kJ	FAT g	SATURATED FAT g	PROTEIN g	CARBO-HYDRATE g	FIBRE g
41	172	2.3	0.9	0.7	4.7	0.1
332	1388	28.4	9.6	19	0	0
491	1809	13	1.6	12	87	2.2
417	1125	30	18.2	31	5	8.4
133	374	9	2.1	4	10	1
213	790	6	Trace	6	37	0
386	1061	27	2.6	0	37	0.8
167	699	1.6	0.3	7.4	32.8	0.5
180	753	5.7	1.1	8.7	25.2	0.7
268	699	21	2.7	9	12	1.3
873	2358	64	10.2	75	7	1
129	539	3.7	0.9	3.7	21.5	0.4
108	451	3	0.8	3.8	17.5	1
60	249	3.9	1.1	3.3	3	0.4
46	194	2	0.5	2	5.4	0
139	583	5.7	1.6	5.1	18	1
99	415	2.2	0.4	3.9	17	0
208	869	7.3	2.4	6.4	31	1
373	1158	20	5	38	11	6.5
248	1046	8.4	1.5	38.2	5.4	1.2
351	882	29	13.7	13	11	9.1
177	455	14	2.4	6	7	6.1
575	1810	30	3.9	23	57	1.8
705	2115	42	6.9	23	63	3
297	912	16	1.8	27	11	4.9
286	1196	8.5	2.3	19.7	35	7.1
158	661	0.4	0.1	7.8	32.8	0.5
153	640	0.7	0.2	8.4	30.1	0.7
150	628	1.3	0.1	3.2	33.6	1.2
125	522	8.7	1.3	2.9	9.3	0.4
155	650	5	1.2	11.7	16.9	0.5

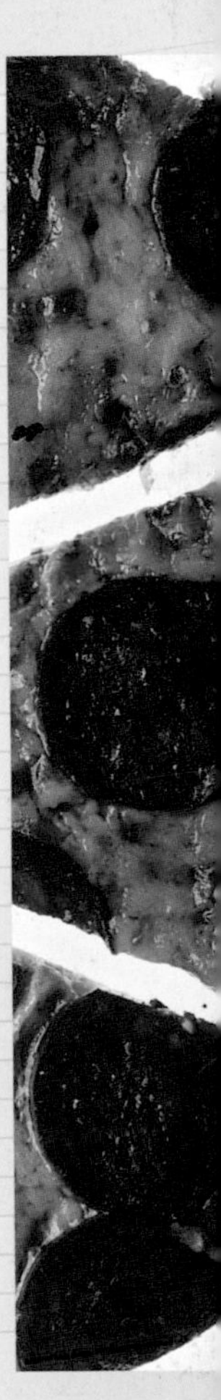

BURGERS AND HOTDOGS	AVERAGE PORTION g	GI	GL
Cheeseburger, large, double	258	H	H
Cheeseburger, large, single	219	H	H
Cheeseburger, regular, double	228	H	H
Cheeseburger, triple, plain	304	H	H
Hamburger, large, double	226	H	H
Hamburger, large, single, with seasoning and sauce	172	H	H
Hamburger, large, triple with, seasoning and sauce	259	H	H
Hamburger, regular, double, plain	176	H	H
Hamburger, regular, double, with seasoning and sauce	215	H	H
Hamburger, regular, single, plain	90	H	H
Hamburger, regular, single, with seasoning and sauce	106	H	H
Hotdog with onions, ketchup and relish	100	H	H

FRIED CHICKEN			
Deep fried chicken breast, boneless	100	H	H
Deep fried chicken drumsticks, bone removed	75	H	H
Deep fried chicken thigh, boneless	100	H	H
Deep fried chicken wing, bone removed	50	H	H
Deep fried coated chicken pieces	100	H	H

MEXICAN DISHES			
Burrito, beans	217	L	L
Burrito, beans and cheese	186	L	L
Burrito, beans and chilli peppers	204	L	L
Burrito, beans and meat	231	L	L
Burrito, beans, cheese and beef	203	L	L
Burrito, beef	220	L	L

Unless otherwise specified, burgers include garnish, seasonings and sauce.

ENERGY kcal	ENERGY kJ	FAT g	SATURATED FAT g	PROTEIN g	CARBO-HYDRATE g	FIBRE g
704	2946	44	17.7	38	40	2.1
563	2354	33	15	28	38	2.1
650	2718	35	12.8	30	53	2.1
796	3332	51	21.7	56	27	2.1
540	2260	27	10.5	34	40	2.1
427	1785	21	7.9	23	37	2.1
692	2893	41	15.9	50	29	2.1
544	2276	28	10.4	30	43	2
576	2410	32	12	32	39	2.1
274	1148	12	4.1	12	31	2
272	1140	10	3.6	12	34	2.3
215	903	9.3	2.3	7.1	27.4	1.8
227	949	11.9	2.7	21.9	8.5	0.4
187	785	11.3	3	15.8	6	0.2
231	969	14.2	3.6	18.6	7.8	0.3
156	652	10.7	2.6	9.8	5.4	0.2
233	972	12.8	3.1	24.8	4.8	0.8
447	1871	13	6.9	14	71	1.5
378	1579	12	6.8	15	55	1.5
412	1724	15	7.6	16	58	1.5
508	2125	18	8.3	22	66	1.5
331	1384	13	7.1	15	40	1.5
524	2191	21	10.5	27	59	1.5

MEXICAN DISHES	AVERAGE PORTION g	GI	GL
Burrito, beef and chilli peppers	201	L	L
Burrito, beef, cheese and chilli peppers	304	L	L
Chimichanga, beef	174	H	H
Chimichanga, beef and cheese	183	H	H
Chimichanga, beef and chilli peppers	190	H	H
Chimichanga, beef, cheese and chilli peppers	180	H	H
Enchilada, cheese	163	M	M
Enchilada, cheese and beef	192	M	M
Enchilada, vegetable	360	M	M
Fajita, beef with vegetables	200	L	L
Fajita, chicken, meat only	170	L	L
Fajita, chicken with vegetables	250	L	L
Nachos, cheese	113	H	H
Nachos, cheese and jalapeno peppers	204	H	H
Nachos, cheese, beans, ground beef and peppers	255	H	H
Tacos/tostada, chicken and vegetables	100	M	H
Tacos/tostada, beef and vegetables	100	M	H

PIZZA			
Pizza, cheese and tomato, deep pan	220	H	H
Pizza, cheese and tomato, thin base	116	H	H
Pizza, fish topped, deep pan	230	H	H
Pizza, fish topped, thin base	150	H	H
Pizza, meat topped, deep pan	230	H	H
Pizza, meat topped, thin base	150	H	H
Pizza, vegetable topped, deep pan	290	H	H
Pizza, vegetable topped, thin base	150	H	H

Unless otherwise specified, pizzas are takeaway.

ENERGY kcal	ENERGY kJ	FAT g	SATURATED FAT g	PROTEIN g	CARBO-HYDRATE g	FIBRE g
426	1783	17	8	22	49	1.5
632	2645	25	10.4	41	64	1.5
425	1777	20	8.5	20	43	1.5
443	1854	23	11.2	20	39	1.5
424	1773	19	8.3	18	46	1.5
364	1521	18	8.4	15	38	1.5
319	1337	19	10.6	10	29	1
323	1350	18	9	12	30	1
525	1690	26	7.6	23	54	7.6
420	1757	15.4	7.8	23.2	50.2	4.2
213	676	11	3.6	29	1	1.7
453	1895	12.8	2.8	24.9	63.4	2.9
438	1208	31	11.6	18	24	3.4
608	2544	34	14	17	60	2
569	2379	31	12.5	20	56	2
178	746	8.9	3	12.8	12.5	1.8
279	1168	16.2	5	13.9	20.7	2.7
547	2310	16	11.4	27	77	4.8
322	1355	12	9.3	17	39	2.2
507	2138	16	6.2	30	64	4.6
343	1446	12	9.2	20	41	2.8
557	2345	21	14	30	67	4.6
391	1642	18	7.5	21	39	2.8
621	2624	18	7.8	32	88	4.9
332	1398	11	4.1	16	44	2.8

RICE, PASTA AND NOODLES	AVERAGE PORTION g	GI	GL
Basmati rice, boiled	230	M	H
Brown rice, boiled	180	M	H
Bulgur wheat, cooked	100	M	M
Cous cous, cooked	100	M	M
Fresh egg pasta, boiled	230	L	H
Macaroni, boiled	230	L	H
Macaroni, wholemeal, boiled	230	L	H
Millet, cooked	100	M	H
Noodles, boiled	230	M	H
Noodles, fried	230	M	H
Noodles, egg, boiled	230	M	H
Noodles, rice boiled	230	M	H
Noodles, soba, boiled	100	L	L
Noodles, udon, boiled	100	M	H
Pasta twists, boiled	230	M	H
Polenta, boiled	100	M	H
Red rice, boiled	230	M	H
Risotto rice, boiled	100	M	H
Quinoa, cooked	100	M	M
Savoury rice, cooked	180	H	H
Semolina, cooked	100	L	H
Short grain/pudding rice, boiled	100	H	H
Spaghetti, boiled	230	L	H
Spaghetti, wholemeal, boiled	230	L	H
White rice, easy cook, boiled	180	H	H
White rice, fried in lard/dripping	300	H	H
White rice, glutinous, boiled	230	H	H
White rice, polished, boiled	230	H	H
Wholemeal pasta, boiled	230	L	H
Wild rice, boiled	100	L	H

Unless otherwise stated, all dishes are homemade.

ENERGY kcal	ENERGY kJ	FAT g	SATURATED FAT g	PROTEIN g	CARBO-HYDRATE g	FIBRE g
405	1698	1.4	0.2	12.9	91.1	1.4
254	1075	2	0.5	5	58	1.4
153	640	0.4	0.1	5.6	33.8	8.2
104	435	0.2	0	3.8	23.2	1.4
350	1467	3.7	0.9	13.3	70.4	2.1
198	840	1	0.2	7	43	2.1
198	840	1	0.2	7	43	6.4
112	467	1	0.2	3.5	23.6	1.3
143	607	1	0.2	6	30	1.6
352	1467	26	1	4	26	1.2
382	1599	2.3	0.5	13.3	82.1	5.1
205	860	0.5	0.2	4.4	49	1.6
102	425	0.1	0	5.1	21.4	1.4
131	548	0.5	0	3.5	30	1
336	1408	0.9	0.2	11	75.7	6.2
40	169	0.2	0	1	9.2	0.9
184	784	1	0.2	4	43	1.4
327	1370	0.4	0.3	7.5	78.3	1.1
139	583	2.2	0.2	5	26.5	2.3
256	1078	6	2	5	47	2.5
104	435	0.3	0.1	4	22.7	1.2
119	498	0.2	0.1	2.4	28.7	0.3
325	1359	1.4	0.5	10.1	72.5	3.5
260	1116	2	0.2	11	53	8.1
248	1057	2	0.5	5	56	0.2
393	1662	10	4.2	7	75	1.8
150	633	1	0.2	4	34	0.5
283	1201	1	0.2	5	68	0.5
308	1289	2.5	0.5	12	63.3	10.1
99	413	0.3	0.1	4	21.3	1.8

RICE AND PASTA DISHES	AVERAGE PORTION g	GI	GL
Cannelloni, meat, shop bought	260	H	H
Cannelloni, spinach	340	H	H
Cannelloni, vegetable	340	H	H
Fresh egg pasta filled with cheese, shop bought, cooked	250	L	M
Fresh egg pasta filled with cheese and tomato, shop bought, cooked	250	L	M
Fresh egg pasta filled with meat, shop bought, cooked	250	L	M
Fresh egg pasta filled with mushrooms, shop bought, cooked	250	L	M
Fresh egg pasta filled with vegetables and cheese, shop bought, cooked	250	L	M
Lasagne, meat, shop bought	420	H	H
Lasagne, spinach	420	H	H
Lasagne, spinach, wholemeal	420	H	H
Lasagne, vegetable	420	H	H
Lasagne, vegetable, wholemeal	420	H	H
Pasta with ham and mushroom sauce	235	H	H
Pasta with meat and tomato sauce	235	H	H
Peppers, stuffed with rice	175	H	H
Pilau, vegetable	180	H	H
Ravioli, canned in tomato sauce	250	H	H
Ravioli, stuffed with cheese, tomato and herbs, shop bought	250	H	H
Ravioli, stuffed with meat and vegetables, shop bought	250	H	H
Rice and blackeye beans	200	H	H
Rice and blackeye beans, brown rice	200	H	H
Risotto, chicken	340	H	H
Risotto, vegetable	290	H	H
Spaghetti bolognese	400	H	H
Spaghetti, canned in tomato sauce	250	H	H
Vine leaves, stuffed with rice	80	M	H

Unless otherwise stated, all dishes are homemade.

ENERGY kcal	ENERGY kJ	FAT g	SATURATED FAT g	PROTEIN g	CARBO-HYDRATE g	FIBRE g
315	1326	13	5.2	17	35	3.1
449	1880	26	7.8	15	43	2.7
493	2067	31	11.6	15	43	2.4
539	2256	16.8	8.8	26.3	75.5	5.3
483	2023	14.8	8.3	21.8	70.3	4
500	2094	11.8	4.5	25.3	78.3	3.5
432	1808	10.8	6	20.3	67.8	4.3
487	2037	14.5	8	21.8	71.8	4
601	2533	26	11.8	31	66	2.9
365	1541	13	5.5	15	53	4.6
391	1659	13	5.5	18	55	9.7
428	1810	18	9.2	17	52	4.2
445	1877	19	9.2	20	52	8.8
284	1194	14	8.2	13	27	2.1
263	1102	10	4	16	30	2
149	630	4	0.7	3	27	2.3
248	1053	8	4.3	5	43	1
193	809	3.5	1.3	5.8	37	4
304	1267	13	7.9	17	30	0.6
324	1350	15	8.9	16	34	3.4
366	1556	7	3	12	68	2.8
350	1488	7	3	11	66	3.6
546	2310	10	4.5	31	84	Trace
426	1708	19	2.9	12	56	6.4
532	2224	24.8	8.6	24.8	49.6	3.6
180	752	0.8	0.3	5.3	40.5	3.3
210	875	14	2.1	2	19	1

BREAD, CAKES, SCONES AND PASTRIES	AVERAGE PORTION g	GI	GL
Asian pastries	40	M	H
Bagel, cinnamon and raisin	100	H	H
Bagel, multigrain	100	H	H
Bagel, plain	100	H	H
Bagel, poppyseed/onion/sesame	100	H	H
Bakewell tarts, iced	50	H	H
Banana loaf, homemade, no icing	85	H	H
Battenberg cake	32	H	H
Brioche	45	H	M
Brown bread, sliced	36	H	L
Brown rolls/baps	48	M	L
Burfi	40	M	H
Cake bar, chocolate	100	H	H
Cake bar, plain	100	H	H
Cake from 'healthy eating' range	24	H	H
Cake with jam and butter cream	60	H	H
Caramel shortcake	50	H	H
Carrot cake, homemade, no icing	85	H	H
Carrot cake, iced, shop bought	100	H	H
Cheese-topped rolls/baps, white	85	H	H
Chelsea buns	78	H	H
Cherry cake	38	H	H
Chinese flaky pastries	40	H	H
Chocolate cake	40	H	H
Chocolate cake with filling and icing	100	H	H
Chocolate fudge cake	98	H	H
Choux buns	112	H	H
Ciabatta, plain	50	H	H
Coconut cake	40	M	H
Cream-filled chocolate pastries	110	H	H
Cream-filled pastries	110	H	H
Cream horns	60	H	H
Crispie cakes	31	H	H
Croissants	60	H	M

ENERGY kcal	ENERGY kJ	FAT g	SATURATED FAT g	PROTEIN g	CARBO-HYDRATE g	FIBRE g
215	897	16	8	3	17	0.8
262	1095	1.7	0.3	9.8	55.2	2.3
254	1063	1.3	0.2	11.3	52.5	7.3
273	1161	1.8	0.4	10	57.8	2.4
269	1125	1.8	0.4	11	55.0	2.4
205	865	8.4	3.5	1.6	32.9	0.6
230	968	9	5.6	4	35	1.5
113	478	3.2	1	1.4	21.1	0.4
144	607	4	2.8	4	24	1
75	317	1	0.2	3	15	1.3
113	482	2	0.3	5	22	1.8
117	487	8	4	5	7	0.8
430	1800	21.6	9.9	5	57.5	1.6
355	1485	12.2	5	3.1	62	0.6
61	262	0.6	0.3	0.8	14.1	0.4
213	896	8.9	4.2	2.2	33.1	0.8
233	974	13.7	7.6	2.4	26.8	0.7
349	1454	25	6.4	5	28	2.1
374	1569	20.2	5.1	4.2	46.8	1.1
241	1016	7	3.4	9	38	1.2
285	1203	11	3.3	6	44	1.3
150	630	6	1.6	2	23	0.4
157	659	7	4	2	24	0.8
182	763	11	8.2	3	20	0.5
413	1730	23.7	9.5	4.5	48.6	1.8
351	1479	14	4.5	5	55	0.9
427	1766	36	19	6	20	0.8
135	574	2	0.3	5	26	1.2
174	726	10	2.6	3	20	1
425	1770	31.4	14.9	6.7	31	1.1
456	1904	30.1	17.7	3.6	45.5	1
261	1082	21	10	2	15	0.5
144	605	6	3.3	2	23	0.1
224	938	12	5.9	5	26	1

BREAD, CAKES, SCONES AND PASTRIES

	AVERAGE PORTION g	GI	GL
Crumpets	40	M	M
Cupcake, chocolate with icing	100	H	H
Cupcake, chocolate, without icing	100	H	H
Cupcake, plain, with icing	100	H	H
Cupcake, plain, without icing	100	H	H
Currant buns	60	H	H
Custard tarts, individual	94	H	H
Danish pastries	110	H	H
Doughnuts, custard-filled	75	H	H
Doughnuts, jam-filled	75	H	H
Doughnuts, ring	60	H	H
Doughnuts, ring, iced	60	H	H
Eccles cakes	45	H	H
Eclairs, fresh	112	H	H
Fancy iced cakes, individual	30	H	H
Focaccia, herb/garlic and coriander	50	H	M
French baguette, white	40	H	M
French baguette, white, part baked	40	H	M
French baton, granary	40	M	M
French stick/flute, white	40	H	M
Fruit buns, white, not iced	78	H	H
Fruit cake	45	M	H
Fruit cake, rich, iced	70	H	H
Fruit cake, wholemeal	90	M	H
Fruit pies, small	54	M	H
Garlic bread	20	H	M
Garlic and herb baguette	40	H	M
Gingerbread	50	H	M
Granary bread, sliced	36	L	L
Granary rolls	56	L	L
Greek pastries	100	H	H
Gulab jamen/gulab jambu	40	H	H
Hot cross buns	50	H	H
Iced buns	45	H	H

ENERGY kcal	ENERGY kJ	FAT g	SATURATED FAT g	PROTEIN g	CARBO-HYDRATE g	FIBRE g
83	352	1	0.2	3	18	0.8
370	1550	14.5	2.9	3.4	60.3	0.8
333	1396	12.5	2.8	5.8	52.7	1.7
355	1485	11.4	2.6	3.1	63.9	0.4
306	1283	9.3	1.6	4.6	54.5	0.8
178	750	5	2	5	32	1.6
247	1036	13.4	5.5	6.3	27	0.9
491	2047	32.1	13	5.3	48	1.7
269	1125	14	10.4	5	32	1.6
241	1014	9.8	3.7	4.1	36.3	1.1
238	997	13	3.8	4	28	1.4
248	1037	15.4	8.5	3.5	25.4	0.8
164	689	8.2	4.2	2	22	0.9
418	1746	27	14.3	5	42	0.6
114	480	5	1.8	1	17.3	0.3
147	620	4	0.6	5	26	0.7
109	465	1	0.1	4	24	1
108	458	1	0.1	4	23	1
110	467	1	0.1	4	22	1
101	431	1	0.1	4	21	1
218	924	4	1.5	6	41	1.7
150	633	5.4	2.1	2	24.8	1.1
249	1053	8	1.8	3	44	1.2
327	1373	14	4.3	5	48	2.2
195	821	7.4	2.6	1.7	32.4	0.9
73	306	4	1.9	2	9	1
139	584	6.7	3.5	2.8	18	0.3
190	799	6	1.9	3	32	0.6
87	369	1	0.3	4	18	1.3
133	565	2	0.3	6	24	2
456	1909	25.1	9.7	6.7	54.1	1.6
143	602	6	3	3	21	0.1
155	657	3	1	4	29	0.9
145	614	3.5	1.6	2.7	27.4	1.4

BREAD, CAKES, SCONES AND PASTRIES	AVERAGE PORTION g	GI	GL
Jam tarts	34	H	H
Jam tarts, wholemeal	34	H	H
Jamaican/golden syrup loaf cake	85	H	H
Jellabi	40	H	H
Macaroons, with chocolate ganache filling	30	H	H
Madeira-type cake, iced	85	M	H
Malt loaf, fruit	35	M	H
Milk bread, white	15	M	H
Mince pies	55	H	H
Muffins, American, chocolate	100	M	M
Muffins, American, fruit/plain	100	M	M
Muffins, English, white	68	M	H
Muffins, English, white, toasted	68	M	H
Naan bread, garlic and coriander/plain	160	H	M
Pitta bread, white	75	M	L
Pumpernickel bread	33	L	L
Rock cakes	29	H	H
Scones, cheese	48	H	H
Scones, fruit	48	H	H
Scones, plain	48	H	H
Scones, potato	57	H	H
Scones, wholemeal	50	H	H
Scones, wholemeal, fruit	50	H	H
Scotch pancakes	31	M	M
Slimmers white bread, sliced	20	M	M
Soda bread, brown	130	M	M
Sourdough bread	100	L	M
Sponge, fatless	58	M	M
Sponge, frozen	60	M	M
Sponge, jam-filled	60	H	H
Sponge, with butter icing	60	M	H
Strawberry tartlets	54	H	H
Swiss roll	30	M	H

ENERGY kcal	ENERGY kJ	FAT g	SATURATED FAT g	PROTEIN g	CARBO-HYDRATE g	FIBRE g
120	505	4.6	1.9	1.1	19.8	0.4
125	528	5	1.6	1	20	1.2
274	1158	9.4	2.7	2.8	47.8	1.3
145	611	5	3	2	24	0.5
151	630	7.8	1.6	3.2	18	0
312	1312	13.4	5.3	4.1	46.7	1.6
18	76	1	0.2	3	22.7	0.9
36	153	1	0.2	1	7	0.3
207	873	8.2	3.3	2.1	33.4	0.9
436	1826	25.4	4.8	5.5	49.5	0.8
375	1569	19.5	2.4	5	47.8	1
152	644	1	0.3	7	30	1.3
177	753	2	0.3	8	35	1.3
456	1929	12	1.6	12	80	3.2
191	813	1	0.7	7	41	1.8
67	283	1	0.1	2	15	2.1
115	483	5	1	2	18	0.4
174	731	9	4.9	5	21	0.8
152	640	5	1.6	4	25	0.8
166	701	5.9	3.1	3.4	26.5	1.1
169	707	8	4.4	3	22	0.9
163	684	7	2.4	4	22	2.6
162	683	6	2	4	24	2.5
91	381	4	1.3	2	14	0.4
46	193	1	0.1	2	9	0.5
267	1132	4	1	11	50	4.9
257	1075	3	0.6	8.8	51.9	3
171	722	4	1	6	31	0.5
190	795	10	3.1	2	24	0.4
181	768	3	1	3	39	1.1
294	1228	18	5.6	3	31	0.4
111	460	6	3.2	1	14	0.6
83	352	1	0.3	2	17	0.2

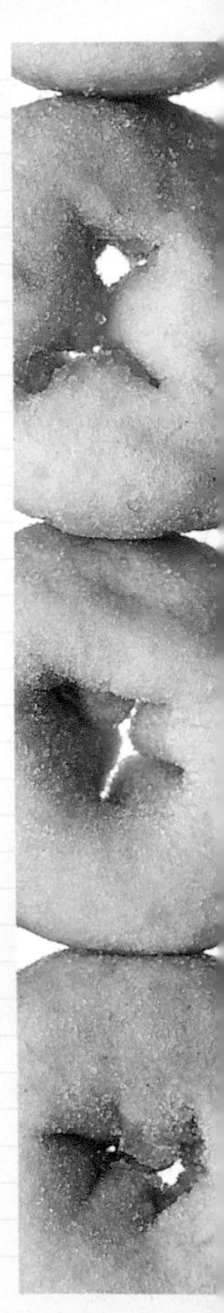

BREAD, CAKES, SCONES AND PASTRIES

	AVERAGE PORTION g	GI	GL
Swiss roll, chocolate	30	M	H
Teacakes	55	H	H
Tortilla, soft	160	M	L
Vanilla slices	113	M	H
Waffles	65	H	H
Wheatgerm bread	30	M	L
White bread, crusty bloomer	35	H	M
White bread, farmhouse, large	35	H	M
White bread, farmhouse, small	27	H	M
White bread, premium	36	H	M
White bread, standard	36	H	M
White bread, standard, toasted	36	H	M
White rolls, crusty	50	H	M
White rolls, soft	45	H	M
Wholemeal bread	38	M	L
Wholemeal bread, toasted	38	M	L
Wholemeal rolls	48	M	L

SANDWICHES

	AVERAGE PORTION g	GI	GL
Bacon, lettuce and tomato, white bread	200	H	H
Cheese and pickle, white bread	200	H	H
Chicken salad, white bread	200	H	H
Egg mayonnaise, white bread	200	H	H
Ham salad, white bread	200	H	H
Prawn cocktail/mayonnaise, white bread	200	H	H
Tuna mayonnaise, white bread	200	H	H

Unless otherwise stated, sandwiches are shop bought.

ENERGY kcal	ENERGY kJ	FAT g	SATURATED FAT g	PROTEIN g	CARBO-HYDRATE g	FIBRE g
124	520	6.8	3.5	1.4	15.3	0.5
181	766	5	1.8	5	32	1.2
451	1903	14	2.9	12	73	3
373	1563	20	10.6	5	45	0.9
217	911	11	4.7	6	26	1
66	281	1	0.2	3	12	1.2
86	368	1	0.2	3	18	0.8
83	353	1	0.2	3	17	0.7
66	281	1	0.2	2	14	0.7
83	352	1	0.2	3	17	0.7
79	335	1	0.1	3	17	0.7
97	414	1	0.1	3	21	0
131	558	1	0.3	5	27	1.2
114	485	1	0.3	4	23	0.9
82	349	1	0.2	4	16	1.9
102	431	1	0.2	5	20	0
117	498	2	0.2	5	22	2.1
470	1966	24.8	5.6	16.4	48.2	2.4
579	2426	29.8	14.8	24	57.4	2.2
351	1467	10.6	2.3	21.4	45.2	2.2
248	1040	12	2.4	8.4	28.5	1.2
334	1399	9	1.9	16.4	50	2.4
469	1963	25.4	2.1	19.2	43.6	4.2
476	1991	21	3.5	24.2	50.6	2

CHEESE	AVERAGE PORTION g	GI	GL
Brie, without rind	40	L	L
Camembert	40	L	L
Cheddar, English, white	40	L	L
Cheddar, half-fat (15% fat)	40	L	L
Cheddar, vegetarian	40	L	L
Cheese spread	30	L	L
Cheshire, white	40	L	L
Cottage (4% fat)	40	L	L
Cottage, low-fat (1.5–2% fat)	40	L	L
Danish Blue	30	L	L
Dolcelatte, without rind	40	L	L
Double Gloucester	40	L	L
Edam	40	L	L
Emmental	40	L	L
Fontina	28	L	L
Fromage frais, fruit	100	L	L
Fromage frais, fruit, virtually fat-free	100	L	L
Fromage frais, natural	100	L	L
Fromage frais, natural, virtually fat-free	100	L	L
Goats' milk	55	L	L
Gouda	40	L	L
Halloumi	40	L	L
Lancashire	40	L	L
Mascarpone	55	L	L
Monterey Jack	28	L	L
Mozzarella, fresh	55	L	L
Mozzarella, grated	55	L	L
Paneer	40	L	L
Parmesan, drums, freshly grated	20	L	L
Parmesan, wedges, freshly grated	20	L	L
Pecorino	28	L	L
Port Salut	40	L	L
Processed cheese, slices	20	L	L
Provolone	28	L	L

Unless otherwise specified, generic cheeses are made with cows' milk.

ENERGY kcal	ENERGY kJ	FAT g	SATURATED FAT g	PROTEIN g	CARBO-HYDRATE g	FIBRE g
144	596	12	7.3	8	2	0
116	482	9	6	9	Trace	0
166	690	14	8.7	10	0	0
109	456	6	4.3	13	Trace	0
156	647	13	8.3	10	Trace	0
81	335	7	4.7	3	2	0
152	630	13	8.5	9	Trace	0
36	149	2	0.9	5	0	0
28	118	1	0.4	5	0	0
103	425	9	5.7	6	Trace	0
158	652	14	8.7	7	Trace	0
165	684	14	9.3	10	Trace	0
136	566	10	6.3	11	Trace	0
160	663	12	8.2	12	Trace	0
110	462	9	5.4	7	0	0
135	566	5.6	0	5.2	16.9	0.4
50	211	0.2	0	6.7	5.6	0.4
113	468	8	5.6	6	4.4	Trace
48	205	0.1	0	7.5	4.6	Trace
174	722	14	9.8	12	Trace	0
151	625	12	8.1	10	Trace	0
124	516	9	6.6	10	0	0
153	633	13	8.4	10	Trace	0
230	949	24	16.2	3	Trace	0
107	443	30	5.4	25	0	0
141	587	11	7.6	10	Trace	0
164	680	12	8.1	14	Trace	0
130	539	10	6.2	10	Trace	0
97	403	7	4.6	9	Trace	0
82	343	6	3.9	7	Trace	0
110	459	8	4.8	1	1	0
133	554	10	7.2	10	Trace	0
59	244	5	2.8	4	1	0
99	417	8	4.8	7	1	0

CHEESE	AVERAGE PORTION g	GI	GL
Red Leceister	40	L	L
Ricotta	55	L	L
Romano	28	L	L
Roquefort	28	L	L
Soft white spreadable, low-fat (<10%)	30	L	L
Soft white spreadable, medium-fat (15%)	30	L	L
Soft white spreadable, full-fat	30	L	L
St Paulin	40	L	L
Stilton, blue	35	L	L
Wensleydale	40	L	L

CREAM AND SUBSTITUTES			
Cream, clotted	45	L	L
Cream, double	30	L	L
Cream, half	30	L	L
Cream, single	15	L	L
Cream, soured	30	L	L
Cream, whipping	30	L	L
Cream, whipping, frozen	30	L	L
Cream, sterilised, canned	45	L	L
Cream, UHT, canned spray	10	L	L
Cream, UHT, half-fat	15	L	L
Cream, UHT, single	15	L	L
Cream, UHT, whipping	30	L	L
Crème fraîche	50	L	L
Crème fraîche, low-fat	50	L	L
Mock cream, double	30	L	L
Mock cream, single	30	L	L
Mock cream, whipping	30	L	L

Unless otherwise specified, generic cheeses are made with cows' milk and cream is fresh.

ENERGY kcal	ENERGY kJ	FAT g	SATURATED FAT g	PROTEIN g	CARBO-HYDRATE g	FIBRE g
161	667	13	8.9	10	Trace	0
79	329	6	3.8	5	1	0
110	459	8	4.8	1	1	0
105	438	9	5.4	6	1	0
43	182	2	1.7	4	1	0
56	231	5	3.4	3	Trace	0
94	386	9	6.7	2	Trace	0
133	554	10	7.2	10	Trace	0
143	594	12	8.1	8	Trace	0
152	632	13	8.5	9	Trace	0
264	1086	29	17.9	1	1	0
135	555	16	9	1	1	0
44	184	4	2.5	1	1	0
30	123	3	1.8	0	1	0
62	254	6	3.8	1	1	0
112	462	12	7.4	1	1	0
114	468	12	7.5	1	1	0
108	443	11	6.7	1	2	0
31	127	2	2	0	0	0
21	86	2	1.2	0	1	0
29	122	3	1.8	0	1	0
112	462	12	7.4	1	1	0
190	792	20	13.2	1.2	1.4	0
85	355	7.5	4.5	1.8	2.5	0
136	561	11	8.7	1	1	0
57	236	4	4.2	1	1	0
96	395	9	8.4	1	1	0

EGGS	AVERAGE PORTION	GI	GL
Boiled	50	L	L
Eggs, raw	60	L	L
Egg white, boiled	27	L	L
Egg white, raw	32	L	L
Egg yolk, boiled	15	L	L
Egg yolk, raw	18	L	L
Duck, boiled and salted	70	L	L
Duck, raw	75	L	L
Fried in sunflower oil	60	L	L
Fried, without fat	50	L	L
Poached	50	L	L
Quail, raw	40	L	L

EGG DISHES			
Omelette, cheese	150	L	L
Omelette, plain (two-egg)	120	L	L
Omelette, Spanish (two-egg)	150	L	L
Scrambled, with milk (two-egg)	120	L	L
Scrambled, without milk (two-egg)	100	L	L

MILK AND MILK SUBSTITUTES			
Coffee whitener	3	L	L
Condensed milk, skimmed, sweetened	15	M	M
Condensed milk, whole, sweetened	15	M	M
Evaporated milk, reduced-fat	15	M	M
Evaporated milk, whole	15	M	H
Flavoured milk	214	M	M
Goats' milk, pasteurised	146	L	L
Semi-skimmed milk, pasteurised	146	L	L
Semi-skimmed milk, UHT	146	L	L
Sheep's milk	146	L	L
Skimmed milk, dried	3	L	L

Unless otherwise specified, figures are for one egg, chicken's eggs are medium-sized and milk is cows' milk.

ENERGY kcal	ENERGY kJ	FAT g	SATURATED FAT g	PROTEIN g	CARBO-HYDRATE g	FIBRE g
79	328	5.4	1.5	7.6	0	0
72	298	4.8	1.4	7.1	0	0
14	60	0	0	3.5	0	0
14	59	0.1	0	3.5	0	0
54	224	4.9	1.4	2.5	0	0
62	259	5.6	1.6	3	0	0
139	575	2.7	11	10	Trace	0
122	510	2.2	9	11	Trace	0
120	499	9.4	2	8.8	0	0
87	363	1.8	6	8	Trace	0
75	309	5.3	1.5	6.7	0	0
60	252	1.2	4	5	Trace	0
399	1659	18.3	34	24	Trace	0
229	950	8.9	20	13	Trace	0
180	752	2.4	12	9	9	2.1
296	1230	13.9	27	13	1	0
160	664	3.3	12	14	Trace	0
16	68	1	1	0	2	0
40	171	0	0	2	9	0
50	211	2	0.9	1	8	0
18	77	1	0.2	1	2	0
23	94	1	0.9	1	1	0
146	614	3	1.9	8	23	0
88	369	5	3.4	5	6	0
67	285	2	1.5	5	7	0
67	283	2	1.6	5	7	0
139	578	8	5.5	8	7	0
10	44	0	0	1	2	0

MILK AND MILK SUBSTITUTES	AVERAGE PORTION g	GI	GL
Skimmed milk, dried, with vegetable fat	3	L	L
Skimmed milk, pasteurised	146	L	L
Skimmed milk, sterilised	136	L	L
Skimmed milk, UHT	146	L	L
Soya milk	146	L	L
Soya milk, flavoured	146	L	L
Whole milk, dried	3	L	L
Whole milk, pasteurised	146	L	L
Whole milk, sterilised	146	L	L
Whole milk, UHT	146	L	L

YOGHURTS			
Crème fraîche (see under cream)			
Fromage frais (see under cheese)			
Drinking yoghurt	200	L	L
Probiotic yoghurt drink, orange	100	M	M
Probiotic yoghurt drink, plain	100	L	L
Yoghurt, French set, fruit, low-fat	125	L	L
Yoghurt, fruit	125	M	M
Yoghurt, fruit, low-fat	125	L	L
Yoghurt, fruit, virtuallly fat-free	125	L	L
Yoghurt, Greek-style, fruit, whole milk	150	M	M
Yoghurt, Greek-style, honey, whole milk	150	M	M
Yoghurt, Greek-style, natural, whole milk	150	L	L
Yoghurt, hazelnut, low-fat	125	L	L
Yoghurt, long life, fruit, whole milk	125	M	M
Yoghurt, natural, low-fat	125	L	L
Yoghurt, natural, virtually fat-free	125	L	L
Yoghurt, soya, fruit	125	M	M
Yoghurt, toffee, low-fat	125	M	M
Yoghurt, twin pot, fruit, virtually fat-free	135	M	M
Yoghurt, vanilla, low-fat	125	L	L

Unless otherwise specified, milk is cows' milk.

ENERGY kcal	ENERGY kJ	FAT g	SATURATED FAT g	PROTEIN g	CARBO-HYDRATE g	FIBRE g
15	61	1	0.5	1	1	0
48	204	1	0.1	5	7	0
44	188	1	0.1	5	7	0
47	200	1	0.1	5	7	0
47	193	2	0.4	4	1	Trace
58	245	2	0.3	4	5	Trace
15	62	1	0.5	1	1	0
96	402	6	3.5	5	7	0
96	404	6	3.5	5	7	0
96	402	6	3.5	5	7	0
124	526	Trace	Trace	6	26	Trace
67	279	0.9	0.6	1.5	13.4	1.3
68	284	1	0.7	1.7	13.1	1.4
103	435	1.4	1	4.3	19.5	0.3
133	564	3.8	2.5	4.9	21.4	0.3
97	412	1.4	0.9	5.1	17.1	0.3
75	318	0.5	0	5.9	12.5	0.3
205	857	12.6	8.4	7.2	16.8	Trace
221	927	12.5	8.4	7.7	21	Trace
198	824	15.3	10.2	8.4	7.2	Trace
111	469	1.9	0.8	5.5	19.1	0.3
125	529	4.3	0	3.9	19.1	0.3
69	294	1.3	0.9	5.9	9.3	Trace
67	286	0.3	0	6.6	10.3	Trace
92	388	2.3	0.3	2.9	16	0.4
114	484	1.1	0.8	4.8	22.6	Trace
55	236	0.1	0.1	4.5	9.7	0.7
114	484	1.1	0.8	4.8	22.6	Trace

FATS	AVERAGE PORTION g	GI	GL
Butter			
Butter	20	L	L
Ghee	7	L	L
Margarine			
Hard block margarine	15	L	L
Soft, animal and vegetable fat	7	L	L
Soft, polyunsaturated	7	L	L
Other fats			
Compound cooking fat	15	L	L
Compound cooking fat, polyunsaturated	15	L	L
Dripping, beef	15	L	L
Ghee, vegetable	7		
Lard	15	L	L
Suet, beef, shredded	15	L	L
Suet, vegetable	15	L	L
Spreads			
Butter, spreadable (75–80% fat)	10	L	L
Butter, spreadable, light (60% fat)	10	L	L
Low-fat spread (26–39%)	7	L	L
Low-fat spread (26–39%) with olive oil	7	L	L
Low-fat spread (26–39%), polyunsaturated	7	L	L
Reduced-fat spread (41–62%)	7	L	L
Reduced-fat spread (41–62%) with olive oil	7	L	L
Reduced-fat spread (41–62%), polyunsaturated	7	L	L
Reduced-fat spread (62–75%)	7	L	L

ENERGY kcal	ENERGY kJ	FAT g	SATURATED FAT g	PROTEIN g	CARBO-HYDRATE g	FIBRE g
147	606	16	11	0	Trace	0
63	259	7	4.6	Trace	Trace	0
103	424	11.5	4	0	0	0
52	213	6	1.9	0	0	0
52	215	6	1.2	Trace	0	0
135	555	15	6.3	0	0	0
135	554	15	3.1	Trace	Trace	0
134	549	15	7.9	Trace	Trace	0
63	259	7	3.3	0	0	0
134	549	15	6	Trace	0	0
124	510	14	7.5	Trace	2	0.1
125	517	13	6.8	0	2	0
72	294	7.9	3.4	0	0.1	0
55	225	6	2.6	0.1	0.1	0
25	106	2.7	0.7	0	0.2	0
25	102	2.7	0.6	0	0	0
23	98	2.6	0.6	0	0.1	0
39	162	4.2	1.1	0	0.1	0
38	157	4.1	0.9	0	0.1	0
37	153	4.1	0.9	0	0	0
46	190	5.1	1.7	0	0	0

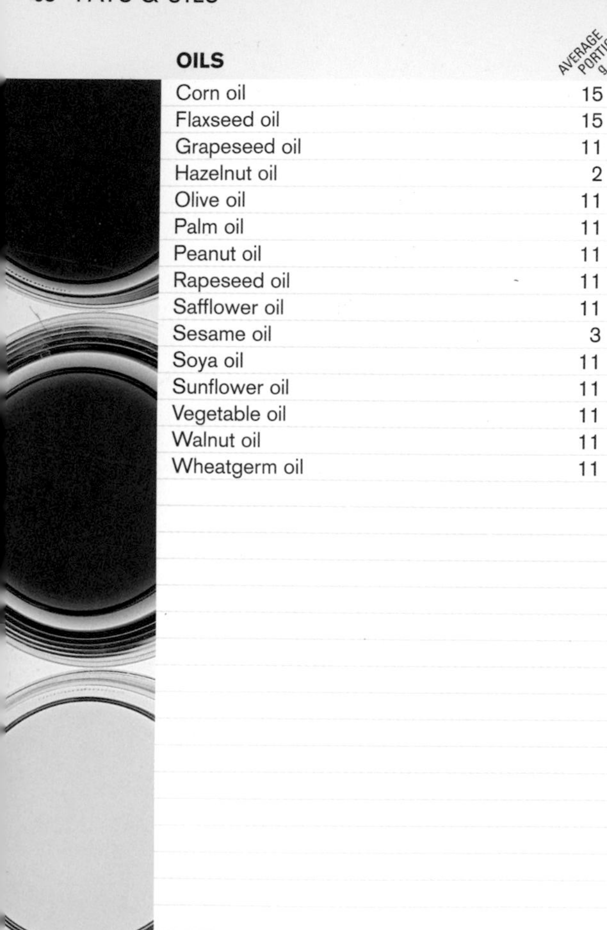

OILS	AVERAGE PORTION g	GI	GL
Corn oil	15	L	L
Flaxseed oil	15	L	L
Grapeseed oil	11	L	L
Hazelnut oil	2	L	L
Olive oil	11	L	L
Palm oil	11	L	L
Peanut oil	11	L	L
Rapeseed oil	11	L	L
Safflower oil	11	L	L
Sesame oil	3	L	L
Soya oil	11	L	L
Sunflower oil	11	L	L
Vegetable oil	11	L	L
Walnut oil	11	L	L
Wheatgerm oil	11	L	L

ENERGY kcal	ENERGY kJ	FAT g	SATURATED FAT g	PROTEIN g	CARBO-HYDRATE g	FIBRE g
135	554	15	2.2	Trace	0	0
135	565	15	1.4	0	0	0
99	407	11	1.2	Trace	0	0
99	407	11	0.9	Trace	0	0
99	407	11	1.6	Trace	0	0
99	407	11	5.3	Trace	0	0
99	407	11	2.2	Trace	0	0
99	407	11	0.7	Trace	0	0
99	407	11	1.1	Trace	0	0
27	111	3	0.4	Trace	0	0
99	407	11	1.7	Trace	0	0
99	407	11	1.3	Trace	0	0
99	407	11	1.1	Trace	0	0
99	407	11	1	Trace	0	0
99	407	11	2	Trace	0	0

SWEET SPREADS	AVERAGE PORTION g	GI	GL
Chocolate spread	35	M	L
Fruit spread	16	M	L
Honey	17	M	L
Lemon curd	15	M	L

SAVOURY SPREADS AND PÂTÉS			
Fish paste	10	L	L
Mackerel pâté, smoked	40	L	L
Meat extract	9	L	L
Meat spread	10	L	L
Liver sausage	40	L	L
Pâté, liver, tubed	40	L	L
Peanut butter, smooth	25	L	L
Peanut butter, crunchy	25	L	L
Sandwich spread	15	L	L
Tofu spread	45	L	L
Tuna pâté	40	L	L
Vegetable pâté	80	L	L
Yeast extract	9	L	L

JAMS AND MARMALADES			
Jam, diabetic	15	L	L
Jam, fruit with edible seeds	15	L	L
Jam, reduced-sugar	15	L	L
Jam, stone fruit	15	L	L
Marmalade	15	L	L
Marmalade, diabetic	15	L	L

Unless otherwise specified, spreads are made with polyunsaturated fats.

ENERGY kcal	ENERGY kJ	FAT g	SATURATED FAT g	PROTEIN g	CARBO-HYDRATE g	FIBRE g
201	841	13.2	2.8	1.2	20.8	0.4
19	83	Trace	Trace	0	5	0.1
49	209	0	0	0	13	0
42	180	1	0.2	0	9	0
17	71	1	0.5	2	0	0
147	608	14	2.5	5	1	Trace
16	68	0	0	4	0	0
19	80	1	0.6	2	0	Trace
90	377	7	2.1	5	2	0.3
114	472	10	3	5	0	Trace
156	645	13	2.0	6	3	1.4
152	628	13	2.4	6	2	1.5
28	117	1	0.2	0	4	0.1
94	349	9	1.3	2	1	0.2
94	393	7	3.1	7	0	Trace
138	574	11	6	6	5	2
16	69	0	0	4	0	0
26	109	0	0	0	9	0.1
39	167	0	0	0	10	0.1
18	78	Trace	Trace	0	5	0.1
39	167	0	0	0	10	0.1
39	167	0	0	0	10	0
26	109	0	0	0	9	0.1

BREAKFAST CEREALS	AVERAGE PORTION g	GI	GL
Bran, flakes	30	H	M
Bran, flakes with oat	30	M	M
Bran, strands	40	L	L
Bran, sultana	30	H	M
Corn flakes	30	H	H
Corn flakes, with nuts	30	H	H
Crunchy clusters	50	H	M
Crunchy clusters with fruit	50	H	M
Crunchy clusters with chocolate	50	H	H
Frosted flakes	30	H	H
Fruit and fibre breakfast cereal	40	H	H
Hoops, honey	30	H	H
Hoops, honey and nut	30	H	H
Malted flakes	30	M	M
Muesli	50	M	L
Muesli, crunchy with nuts	50	M	L
Muesli, Swiss-style	50	M	L
Muesli, with extra fruit	50	M	L
Muesli, with no added sugar	50	M	L
Multigrain flakes	30	M	M
Oat cereal with fruit and nuts	50	M	M
Oat cereal with tropical fruit	50	M	M
Oat clusters	30	M	M
Oat flakes	30	M	M
Porridge, instant, made with water	180	L	M
Porridge, made with water	160	L	L
Porridge, made with whole milk	160	L	L
Puffed wheat	20	H	H
Rice pops	30	H	H
Rice pops, chocolate	30	H	H
Wheat, shredded	45	M	M
Wheat, shredded, honey and nut	40	M	M
Wheat, shredded, mini	45	M	M
Wholewheat biscuits	20	M	M

All values are without any sugar added.

ENERGY kcal	ENERGY kJ	FAT g	SATURATED FAT g	PROTEIN g	CARBO-HYDRATE g	FIBRE g
95	406	1	0.1	3	21	3.9
105	435	2	0.2	3	20	3
104	444	1	0.2	6	19	9.8
96	405	1	0.1	3	18	3.9
108	461	Trace	0	2	26	0.3
119	507	1	0.2	2	27	0.2
200	844	5.8	2.1	3.6	35.5	2.2
139	580	5	1.7	4	20	1.3
220	917	8	2.7	3	33	2.4
113	482	Trace	0	2	28	0.2
147	614	2	1.2	2	28	2.2
111	465	1	0.2	2	23	2.1
112	474	1	0.3	2	23	1.5
186	773	1	0.1	3	23	1.3
184	770	3	0.7	6	33	3.2
225	946	10.3	2.3	4.2	31	2.2
182	770	3	0.4	5	36	3.2
186	789	3	0.4	5	37	3.2
183	776	4	0.8	5	34	3.8
111	465	Trace	Trace	5	22	0.7
213	888	8	3	5	31	2
217	905	7	4	4	34	3
116	490	3	0.8	3	20	2.7
107	456	1	0.2	3	22	3
671	2844	14	2.2	21	123	13
78	334	2	0.3	2	14	1.3
186	781	8	4.3	8	22	1.3
64	273	Trace	0	3	13	1.1
111	472	0.1	0.1	2	27	0.2
115	491	0.1	0.1	2	28	0.2
150	630	1	0.2	5	30	5.2
152	642	3	0.9	4	28	4.1
154	655	1	0.2	4	32	5
70	300	0.1	0.1	2	15	1.9

SAVOURY BISCUITS	AVERAGE PORTION g	GI	GL
Breadsticks	10	M	M
Cheese-flavoured biscuits	14	M	M
Cheese sandwiches	24	M	M
Cheese straws/twists	24	M	M
Cream crackers	16	M	M
Crispbakes	16	M	M
French toast	16	M	M
Matzos	20	M	M
Melba toast	6	M	M
Oatcakes	20	L	L
Rice cakes	16	H	H
Rye crispbread	37	M	M
Toasted minibreads	15	H	M
Water biscuits	16	M	M
Wholemeal crackers	20	M	M

SWEET BISCUITS			
All-butter biscuits	19	H	M
Bourbon creams	26	H	M
Brandy snaps	30	H	M
Cereal bars, with fruit and/or nuts	28	M	M
Cereal bars, with fruit and/or nuts and chocolate	28	M	M
Chocolate-coated biscuits	50	H	H
Chocolate-coated biscuits with marshmallow	22	H	H
Chocolate-coated chocolate wafer biscuits	24	H	H
Chocolate-coated cream-filled biscuits	20	H	H
Chocolate-coated shortcake biscuits	26	H	H
Chocolate-coated teacakes	26	H	H
Coconut biscuits	40	H	H
Cookies, chocolate chip, standard	22	H	M

Unless otherwise specified, a helping is two biscuits.

ENERGY kcal	ENERGY kJ	FAT g	SATURATED FAT g	PROTEIN g	CARBO-HYDRATE g	FIBRE g
39	165	0.8	0.6	1.1	7.3	0.2
69	290	3.9	1.6	1.5	7.4	0.3
123	516	5	3.5	1	17	0.2
122	512	7.3	4.2	3.4	11.6	0.6
71	300	2.6	1.2	1.4	11.2	0.5
60	250	Trace	Trace	2	12	0.6
63	262	1	0.5	2	12	0.5
77	327	0	0	2	17	0.6
23	94	Trace	Trace	1	4	0.3
91	381	4	1.1	1.9	12.6	1.8
56	234	Trace	Trace	2	12	0.8
105	448	0.5	0.1	3.2	23.5	5.3
63	268	2	0.5	1.6	10.3	0.4
70	297	1	Trace	2	12	0.5
83	349	2	1	2	14	0.9
90	378	4	3.4	Trace	14	0.2
124	520	6.1	3.5	1.3	17.1	0.6
131	550	6	3	1	19	0.2
99	419	3	1.1	1.6	17.6	1
122	514	5.1	2.4	1.7	18.5	0.8
262	1099	13	8.4	3	34	1
91	382	4.2	2.3	1	13.2	0.4
122	511	6.9	4.2	1.3	14.7	0.3
99	415	5.6	3.2	1	11.8	0.4
132	551	7.1	3.9	1.7	16.3	0.6
116	438	4	2.2	2	16	0.4
202	846	12	6	2	21	2
104	434	5.5	2.7	1.2	13.2	0.5

SWEET BISCUITS	AVERAGE PORTION g	GI	GL
Cookies, chocolate chip, American-style	28	H	M
Cookies, chocolate chip and hazelnut (1)	24	H	M
Cookies, Danish butter	22	H	M
Cookies, stem ginger	32	H	M
Crunch biscuits	27	H	M
Custard creams	26	H	M
Custard creams, reduced fat	24	H	M
Digestives, half coated in chocolate	26	H	M
Digestives, plain	26	H	M
Digestives with oats, plain	26	M	M
Digestives with oats, reduced fat	30	M	M
Fig rolls	18	H	M
Flapjacks (1)	70	H	M
Flapjacks, shop bought	50	H	M
Fruit/Garibaldi biscuits	20	H	M
Gingernuts	20	H	M
Ice cream wafers, plain not filled	10	H	M
Iced biscuits	26	H	M
Jaffa cakes	26	H	M
Jam-filled rings	30	H	M
Jam sandwich creams	30	M	M
Malted milk/shortcake-type biscuits	20	H	M
Malted milk creams	26	H	M
Marshmallow teacakes, chocolate-coated	28	H	M
Nice	20	H	M
Nice creams	28	H	M
Rich tea biscuits	10	H	M
Rich tea biscuits, half coated in chocolate	26	H	M
Shortbread	26	H	M
Shortbread, chocolate chip	40	H	M
Snowballs	42	H	M
Wafers, filled	14	H	M

Unless otherwise specified, a helping is two biscuits.

ENERGY kcal	ENERGY kJ	FAT g	SATURATED FAT g	PROTEIN g	CARBO-HYDRATE g	FIBRE g
123	517	6	2.7	1.5	17	0.5
120	504	6	2.6	2	14	0.4
112	461	5	4	1	14	0.3
157	653	9	1.9	1	20	0.6
135	562	7	3.5	2	17	0.6
124	520	6.1	3.5	1.3	17.1	0.6
114	472	4	2.6	2	18	0.4
127	532	6.7	3.3	1.7	16.1	0.8
120	505	5.5	2	1.6	17.1	0.7
125	524	6	1.5	1.7	17.3	1.1
130	547	4.1	1.3	2	22.7	0.8
65	276	1.9	0.9	0.7	12	0.6
339	1420	19	5.3	3	42	1.9
217	910	11.4	5.1	2.6	27.8	1.1
82	347	3.3	1.5	1	13.1	0.5
89	373	3.1	1.5	1	15.1	0.3
37	156	0.3	0.1	1	8	0.2
105	446	2.8	1.3	1.3	20	0.4
94	398	3	0.6	1	18	0.
128	542	4.3	2	1.6	22.2	0.6
142	592	6	3.2	2	20	0.6
96	403	4.2	2	1.1	14.3	0.3
130	542	6	3.7	2	17	0.4
126	525	5	3	1	18	0.4
94	393	4	2.3	1	12	0.3
140	584	6	3.2	2	18	0.6
44	187	1.5	0.5	0.6	7.5	0.2
132	552	6.3	3.3	1.6	18.3	0.5
134	561	7.5	4.5	1.4	16.2	0.3
206	858	10	8.6	2	24	0.6
106	814	10	9.8	2	24	2.2
75	314	4	2.6	1	9	0.3

SAVOURY SNACKS	AVERAGE PORTION g	GI	GL
Bacon streaks	25	H	M
Bombay mix	30	M	M
Breadsticks	28	M	M
Cheese and potato puffs	27	H	M
Cheese balls	33	H	M
Corn snacks	28	H	M
Oriental mix	50	H	M
Popcorn, candied	75	H	M
Popcorn, salted	75	L	L
Pork scratchings	22	M	M
Potato and corn sticks	19	H	M
Potato and corn, waffle-shaped snacks	50	H	M
Potato crisps	30	H	M
Potato crisps, crinkle cut	40	H	M
Potato crisps, fried in sunflower oil	30	H	M
Potato crisps, jacket	40	H	M
Potato crisps, low-fat	30	H	M
Potato crisps, square	25	H	M
Potato crisps, thick, crinkle cut	40	H	M
Potato crisps, thick cut	40	H	M
Potato rings	30	H	M
Prawn crackers	20	H	M
Pretzels	30	H	M
Punjabi puri	35	H	M
Tortilla chips	50	H	M
Trail mix	25	M	M
Twiglets	25	H	M
Vegetable crisps	40	H	M
Wheat crunchies	35	H	M

ENERGY kcal	ENERGY kJ	FAT g	SATURATED FAT g	PROTEIN g	CARBO-HYDRATE g	FIBRE g
113	473	5	1.6	2	16	0.5
151	630	10	1.2	6	11	1.9
108	464	4	1.6	4	20	0.8
140	585	9	3.2	2	15	0.3
148	621	8	1.8	2	19	0.8
147	616	8.5	0.8	1.7	17	0.4
273	1142	20	3	10	17	0.3
360	1514	14	1.5	2	58	1
445	1851	32	3.2	5	37	1
133	554	10	3.6	11	0	0.1
88	368	4	1.3	1	11	0.6
241	1009	12	2.8	2	32	1.3
156	650	9.5	2.5	1.3	17.2	0.7
219	913	14	5.8	2	22	2.3
150	629	8.9	0.9	1.8	16.7	1.4
204	851	13	5.3	3	21	1.9
137	577	6	2.8	2	19	1.8
108	454	5	2.2	2	14	1.1
203	848	12	5	2	22	1.6
200	836	11	4.6	3	23	1.6
144	605	6.7	0.6	1.1	21.2	0.5
103	432	7	1.9	0	11	0.3
114	479	1	0.2	3	24	0.8
188	784	12	2.9	3	18	0.9
252	1055	13.7	1.4	3.6	30.4	2.9
108	451	7	1.3	2	9	1.1
96	404	3	1.2	3	16	2.6
192	804	13.9	1.6	1.9	15.8	5.0
158	662	7	3.2	4	21	1.2

FLOUR	AVERAGE PORTION g	GI	GL
Chapati, brown	100	H	H
Chapati, white	100	M	H
Chickpea	100	H	H
Cornflour	30	H	M
Millet	100	L	H
Rice	100	H	H
Rye, whole	20	M	H
Wheat, brown	100	H	H
Wheat, white, breadmaking	100	H	H
Wheat, white, plain	100	H	H
Wheat, white, self-raising	100	H	H
Wheat, wholemeal	100	M	H

GRAVY AND STOCK CUBES			
Instant granules	5	L	L
Instant granules, made up	45	L	L
Stock cubes, beef/chicken	7	L	L
Stock cubes, vegetable	7	L	L

MISCELLANEOUS			
Artichoke hearts, bottled in oil, drained	50	L	L
Baking powder	4	L	L
Bran, wheat	7	M	L
Capers in brine, drained	10	L	L
Chocolate chips, dark	100	H	H
Chocolate chips, milk	100	H	H
Chocolate chips, white	100	H	H
Coconut, creamed block	25	L	L
Coconut, desiccated	28	L	L
Coconut milk	100	L	L
Coconut milk, light	100	L	L
Custard powder	30	L	L

Unless otherwise specified, all gravies and stocks do not contain flour.

ENERGY kcal	ENERGY kJ	FAT g	SATURATED FAT g	PROTEIN g	CARBO-HYDRATE g	FIBRE g
333	1419	1	0.2	12	74	6
335	1426	1	0.1	10	78	3
313	1328	5.4	0.5	19.7	49.6	10.7
106	452	0	0	0	28	0
354	1481	1.7	0.7	5.8	75.4	8.5
366	1531	0.8	0.1	6.4	80.1	2
67	286	1	0.1	2	15	2.3
323	1377	2	0.2	13	69	6.4
341	1451	1	0.2	12	75	3.1
341	1450	1	0.2	9	78	3.1
330	1407	1	0.2	9	76	3.1
310	1318	2	0.3	13	64	9
23	96	2	Trace	0	2	Trace
15	64	1	Trace	0	1	Trace
17	69	1	Trace	1	1	0
18	74	1	Trace	1	1	Trace
63	265	4.2	0.6	1.6	5.1	2.5
7	28	0	0	0	2	0
14	61	1	0.1	1	2	2.5
4	15	0.1	0	0.2	0.5	0.3
496	2075	30	17.7	4.1	55.8	1.3
502	2101	27.6	14.4	5.6	61.6	0.8
531	2225	32.1	28.6	7.1	57.1	0
167	690	17	14.8	2	2	2
169	698	17	15	2	2	3.8
153	642	15	13.3	1.4	3.4	0.1
72	301	7	6.2	0.7	1.6	0.2
106	452	0	0	0	28	0

MISCELLANEOUS	AVERAGE PORTION g	GI	GL
Gelatine	3	L	L
Marzipan, shop bought	25	H	H
Mincemeat	28	H	H
Mincemeat, vegetarian	28	H	H
Olives, in brine, stoned	18	L	L
Olives, green, stuffed	40	L	L
Olives, mixed	40	L	L
Olives, mixed with feta	50	L	L
Salt	5	L	L
Thai red curry paste	15	L	L
Thai green curry paste	15	L	L
Tomatoes, sundried, bottled in oil	10	L	L
Vinegar	15	L	L
Wheatgerm	5	M	L
Yeast, bakers', compressed	5	L	L
Yeast, dried	5	L	L

NUTS AND SEEDS			
Almonds, toasted	13	L	L
Barcelona nuts	15	L	L
Brazil nuts	10	L	L
Caraway seeds	10	L	L
Cashew nuts, plain	10	L	L
Cashew nuts, roasted and salted	25	L	L
Chestnuts	50	L	L
Chia seeds	10	L	L
Coconut, fresh	28	L	L
Flaxseeds	10	L	L
Hazelnuts	10	L	L
Hemp seeds	10	L	L
Macadamia nuts, salted	10	L	L
Melon seeds	15	L	L
Mixed nuts	40	L	L

ENERGY kcal	ENERGY kJ	FAT g	SATURATED FAT g	PROTEIN g	CARBO-HYDRATE g	FIBRE g
10	43	0	0	3	0	0
101	426	3	0.3	1	17	0.5
77	326	1	Trace	0	17	0.4
85	359	4	Trace	0	12	0.4
19	76	2	0.3	0	Trace	0.5
52	219	5	0.7	0.4	1.5	1.2
45	188	3.8	0.5	0.4	2.4	1.2
78	326	7.5	0.7	2.2	0.4	0.3
0	0	0	0	0	0	0
25	103	2.1	0.1	0.1	1.4	0.2
17	71	1.2	0.5	0.3	1.3	0.5
50	204	5	0.7	0	1	0.5
3	13	0	0	0	0	0
18	75	0	0.1	1	2	0.8
3	11	0	0	1	0	Trace
8	36	0	0	2	0	Trace
81	334	7	0.6	3	1	1
96	396	10	2.3	2	1	1
68	281	7	1.6	1	0	0.4
40	169	1.5	0.1	2	5	3.8
57	237	5	1	2	2	0.3
153	633	13	2.5	5	5	0.8
85	360	1	0.3	1	18	2
51	213	3.1	0.3	1.6	4.4	3.8
98	405	10	8.7	1	1	2
56	234	4.2	0.4	1.8	2.0	2.7
65	269	6	0.5	1	1	0.7
60	250	4.7	0.3	3.7	0.7	0.2
75	308	8	1.1	1	0	0.5
87	363	7	1.8	4	1	0.8
243	1006	22	3.4	9	3	2.4

NUTS AND SEEDS	AVERAGE PORTION g	GI	GL
Mixed nuts and raisins	40	M	M
Peanuts, dry roasted	40	L	L
Peanuts, plain	13	L	L
Peanuts, roasted and salted	25	L	L
Pecan nuts	60	L	L
Pine nuts	5	L	L
Pistachio nuts, roasted and salted	10	L	L
Poppy seeds	10	L	L
Pumpkin seeds	16	L	L
Sesame seeds	12	L	L
Sunflower seeds	16	L	L
Walnuts	20	L	L

PURÉES			
Tomato purée	15	L	L
Vegetable purée	20	L	L

SUGARS, SYRUPS AND TREACLE			
Jaggery	16	M	M
Sugar, brown	20	M	M
Sugar, Demerara	20	M	M
Sugar, icing	20	M	M
Sugar, white	20	M	M
Syrup, golden	55	M	M
Syrup, golden, pouring	50	M	M
Syrup, maple	20	M	M
Treacle, black	50	M	M

ENERGY kcal	ENERGY kJ	FAT g	SATURATED FAT g	PROTEIN g	CARBO-HYDRATE g	FIBRE g
192	802	14	2.2	6	13	1.8
236	976	20	3.6	10	4	2.6
73	304	6	1.1	3	2	0.8
151	623	13	2.4	6	2	1.5
413	1706	42	3.4	6	3	2.8
34	142	3	0.2	1	0	0.1
60	249	6	0.7	2	1	0.6
57	237	4.5	0.5	1.8	2.4	1
91	378	7	1.1	4	2	0.8
72	296	7	1	2	0	0.9
96	400	8	0.8	3	3	1
138	567	14	1.1	3	1	0.7
10	43	0	0	0.7	1.9	0.5
14	51	1	0	1	1	0.6
59	250	0	0	0	16	0
72	309	0	0	0	20	0
79	336	0	0	0	21	0
79	336	0	0	Trace	21	0
79	336	0	0	Trace	21	0
164	698	0	0	0	43	0
148	632	0	0	Trace	40	0
52	219	0	0	0	13.4	0
129	548	0	0	1	34	Trace

TABLE SAUCES

	AVERAGE PORTION g	GI	GL
Apple sauce, homemade	20	M	L
Balsamic vinegar	15	L	L
Brown sauce	20	M	L
Brown sauce, hot	20	M	L
Brown sauce, sweet	20	M	L
Chilli sauce	25	M	L
Horseradish sauce	20	M	L
Mint sauce	10	M	L
Mustard, powder, made up	8	L	L
Mustard, smooth	2	L	L
Mustard, wholegrain	14	L	L
Soy sauce, dark, thick	5	M	L
Soy sauce, light, thin	5	M	L
Tartare sauce	30	M	L
Tomato ketchup	20	M	L
Worcestershire sauce	10	L	L

WHITE SAUCES

Bread sauce, made with semi-skimmed milk	45	M	M
Bread sauce, made with whole milk	45	M	M
Cheese sauce, made with semi-skimmed milk	62	M	M
Cheese sauce, made with whole milk	62	M	M
Hollandaise, homemade	30	M	M
Hollandaise, shop bought	30	M	M
Onion sauce, made with semi-skimmed milk	62	M	M
Onion sauce, made with whole milk	62	M	M
White sauce, savoury, made with semi-skimmed milk	62	M	M
White sauce, savoury, made with whole milk	62	M	M

ENERGY kcal	ENERGY kJ	FAT g	SATURATED FAT g	PROTEIN g	CARBO-HYDRATE g	FIBRE g
13	55	0	0	0	3	0.2
15	63	0	0	0	4	0
20	84	Trace	Trace	0	5	0.1
24	102	Trace	Trace	0	6	0.1
20	84	Trace	Trace	0	4	0.1
20	84	Trace	Trace	0	4	0.3
31	128	2	0.2	1	4	0.5
10	43	Trace	Trace	0	2	0.2
18	75	1	0.1	1	0	0
3	12	0	0	0	0	0
20	82	1	0.1	1	5	Trace
3	13	0	0	0	6	0.2
3	13	Trace	Trace	0	1	0.7
90	372	7	0.5	0	0	0.2
23	98	Trace	Trace	0	1	0.7
6	28	0	0	0.1	1.6	0
42	177	1	0.6	2	6	0.1
50	208	2	1.2	2	6	0.1
112	465	8	3.9	4	6	0.1
122	508	8	4.8	4	6	0.1
215	900	23.1	13.8	1.5	0.3	0
136	570	14.2	8.3	1.5	0.6	0
53	224	3	1.1	2	5	0.2
61	257	4	1.7	2	5	0.2
79	334	5	1.8	3	7	0.1
93	387	6	2.7	3	7	0.1

WHITE SAUCES	AVERAGE PORTION g	GI	GL
White sauce, sweet, made with semi-skimmed milk	62	M	M
White sauce, sweet, made with whole milk	62	M	M

CHUTNEYS			
Chutney, apple, homemade	33	M	M
Chutney, mango, oily	33	M	M
Chutney, mango, sweet	33	M	M
Chutney, mixed fruit	33	M	M
Chutney, tomato	33	M	M

DIPS			
Dips, sour-cream-based	30	L	L
Guacamole	45	L	L
Hummus	30	L	L
Salsa	30	L	L
Taramasalata	45	L	L
Tzatziki	45	L	L

DRESSINGS			
Blue cheese	25	M	M
Caesar	15	M	M
'Fat-free'	15	L	L
French	15	L	L
Honey mustard	15	M	M
Italian salad	15	L	L
Low-fat	15	L	L
Oil and lemon	15	L	L
Ranch salad	15	M	M
Thousand island	30	L	L

ENERGY kcal	ENERGY kJ	FAT g	SATURATED FAT g	PROTEIN g	CARBO-HYDRATE g	FIBRE g
93	393	4	1.7	2	12	0.1
105	441	6	2.5	2	12	0.1
66	283	Trace	Trace	0	17	0.4
94	397	4	Trace	0	16	0.3
62	266	Trace	Trace	0	16	0.3
51	219	Trace	Trace	0	13	0.3
42	179	Trace	Trace	0	10	0.4
108	445	11	3.7	1	1	Trace
58	239	6	1.2	1	1	1.1
56	234	4	0.5	2	3	0.7
20	86	1.1	0	0.3	2.5	0.4
227	935	24	1.8	1	2	Trace
30	124	2	1.3	2	1	0.1
114	472	12	6.2	1	2	0
65	271	6.9	1.3	0.2	0.5	0
10	42	0	0	0	2	0
69	285	7	0.6	0	1	0
49	205	2.6	0.3	0.1	6.7	0.1
45	189	4.3	0.7	0.1	1.6	0
11	45	1	0.1	0	1	Trace
97	399	11	1.1	0	0	Trace
74	309	7.7	1.4	0.2	1	0.1
97	401	9	0.9	0	4	0.1

DRESSINGS	AVERAGE PORTION g	GI	GL
Thousand island, reduced-calorie	30	L	L
Yoghurt-based	30	L	L
Mayonnaise, homemade, made with lemon juice	30	L	L
Mayonnaise, homemade, made with vinegar	30	L	L
Mayonnaise, reduced-calorie	30	L	L
Mayonnaise, shop bought	30	L	L
Salad cream	20	L	L
Salad cream, reduced-calorie	20	L	L

PICKLES			
Gherkins, pickled	25	L	L
Onions, pickled	30	L	L
Piccalilli	40	L	L
Pickle, chilli, oily	15	L	L
Pickle, chow chow, sour	15	L	L
Pickle, chow chow, sweet	15	L	L
Pickle, lime, oily	15	L	L
Pickle, mango, oily	15	L	L
Pickle, mixed vegetables	15	L	L
Pickle, sweet	15	L	L

COOKING SAUCES			
Barbecue sauce	25	M	L
Black bean sauce	20	M	L
Bolognese sauce, fresh	100	L	M
Carbonara pasta sauce, fresh	100	L	M
Chilli sauce	25	M	L
Chinese cook-in sauce, sweet and sour	140	M	M
Chinese stir fry sauce	140	M	M

ENERGY kcal	ENERGY kJ	FAT g	SATURATED FAT g	PROTEIN g	CARBO-HYDRATE g	FIBRE g
59	243	5	0.5	0	4	Trace
88	363	8	0.8	1	3	Trace
237	974	26	3.8	1	0	Trace
217	894	24	3.5	1	0	0
86	356	8	1.1	0	2	0
206	847	22.4	1.7	0.3	0.7	0
70	288	6	0.8	0	3	0
39	161	3	0.5	0	2	0
4	15	Trace	Trace	0	1	0.3
7	30	Trace	Trace	0	1	0.4
34	144	0	0	0	7	0.4
41	168	4	Trace	0	1	0.2
4	18	0	0	0	1	0.2
17	73	0	0	0	4	0.2
27	111	2	Trace	0	1	0.2
27	110	2	Trace	0	1	0.2
3	14	0	0	0	1	0.2
21	91	0	Trace	0	5	0.2
23	99	0	0	0	6	0.1
19	79	0	0	1	2	0.4
44	182	1.3	0.2	1.5	6.9	1.3
151	632	13	7.7	4.9	3.8	0.4
20	84	Trace	Trace	0	4	0.3
131	550	1.4	0.2	0.7	30.9	1.3
129	539	1.8	0.2	2	27.9	1.1

COOKING SAUCES	AVERAGE PORTION g	GI	GL
Curry sauce, canned	150	M	L
Fish sauce	10	L	L
Four cheese pasta sauce, fresh	100	M	M
Indian cook-in sauce, korma/tikka masala	140	M	M
Indian cook-in sauce, tomato based	140	M	M
Oyster sauce	15	L	L
Pesto, green	100	L	M
Pesto, red	100	L	M
Sauce, curry, sweet	150	M	M
Sauce, curry, tomato and onion	150	M	M
Soy sauce, dark, thick	5	L	L
Soy sauce, light, thin	5	L	L
Tomato-based cook-in sauce	140	L	M
Tomato sauce, fresh	100	L	M
Tomato sauce with vegetables, fresh	100	L	M
Pasta sauce, tomato based	170	L	M
White milk/cream-based cook-in sauce	140	M	M

CUSTARDS			
Crème caramel	90	L	L
Custard, canned, ready-to-serve	120	L	L
Custard, canned, ready-to-serve, low-fat	120	L	L
Custard, chilled, ready-to-serve	120	L	L
Custard, confectioners'	150	L	L
Custard, made with semi-skimmed milk	150	L	L
Custard, made with whole milk	120	L	L

ENERGY kcal	ENERGY kJ	FAT g	SATURATED FAT g	PROTEIN g	CARBO-HYDRATE g	FIBRE g
117	486	8	0.4	2	11	Trace
4	15	0	0	0.5	0.4	0
164	685	12.3	7.3	5.1	8.7	0.9
186	781	13.3	6.1	2.4	15.3	2.1
94	393	4.5	0.7	2	12.2	1.8
12	51	0	Trace	1	3	Trace
412	1725	42.5	6.5	3.7	3.9	1.1
307	1284	30.6	4.7	5	3	1
137	570	8	2.3	2	14	2.1
297	1229	29	3	3	9	1.7
3	13	0	0	0	0	00
3	13	0	0	0	0	0
60	250	1.4	0.2	1.3	11.2	1.7
33	140	0.3	0.1	1.4	6.7	1.8
46	192	2.1	0.3	1.4	5.7	2.4
80	340	3	0.3	3	12	2.5
147	614	12	2.4	2.2	7.8	2.7
98	416	1	Trace	3	19	Trace
118	497	3	2.3	3	20	0
93	395	2	0	3	18	0
135	569	6	4.4	3	18	0
255	1077	9	4.1	10	37	0.3
141	605	3	1.8	6	25	Trace
140	594	5	3.4	4	20	Trace

PASTRY	AVERAGE PORTION g	GI	GL
Cheese	100	H	H
Choux	100	H	H
Filo, cooked	100	H	H
Filo, uncooked	100	H	H
Flaky/puff, cooked	100	H	H
Flaky/puff, uncooked	100	H	H
Shortcrust, cooked	100	H	H
Shortcrust, uncooked	100	H	H
Wholemeal	100	H	H
PASTRY DISHES			
Beef pie with pastry, individual	150	H	H
Chicken and mushroom pie, single crust	100	L	L
Chicken pie, individual, baked	130	L	L
Cornish pasty	155	H	H
Cornish pasty, shop bought	155	H	H
Lamb samosa, baked	70	H	H
Lamb samosa, deep-fried	70	H	H
Pasty, vegetable	155	L	H
Pasty, vegetable, wholemeal	155	L	H
Pork and egg pie	60	H	H
Pork pie	60	H	H
Pork pie, mini	50	H	H
Quiche, broccoli	140	L	H
Quiche, broccoli, wholemeal	140	L	H
Quiche, cauliflower cheese	140	L	H
Quiche, cauliflower cheese, wholemeal	140	L	H
Quiche, cheese and egg	140	L	H
Quiche, cheese and egg, wholemeal	140	L	H
Quiche, cheese and mushroom	140	L	H
Quiche, cheese and mushroom, wholemeal	140	L	H

Unless otherwise specified, all pastry dishes are homemade.

ENERGY kcal	ENERGY kJ	FAT g	SATURATED FAT g	PROTEIN g	CARBO-HYDRATE g	FIBRE g
500	2083	34	15.3	13	37	1.5
325	1355	20	15	9	30	1.2
363	1544	3.8	0.4	10	77.1	2.2
278	1180	2.9	0.3	7.6	58.9	1.7
486	2027	33.2	15.9	6.7	42.8	1.1
384	1600	26.2	12.6	5.3	33.7	0.9
547	2281	37.9	14.1	6.9	47.5	2.2
453	1889	31.4	11.7	5.7	39.4	1.8
499	2080	33	11.8	9	45	6.3
438	1830	26.6	11.9	13.8	38.3	2
200	836	10	4.5	13	14	0.6
374	1563	21	9.1	12	32	1
431	1800	27.6	13.2	10.9	37.2	1.7
414	1731	25	9.1	10	39	1.4
187	781	10	2.5	8	16	1.2
265	1097	22	3.3	6	12	0.8
425	1783	23	5.7	6	52	2.9
406	1702	24	5.7	8	43	6.4
178	740	13	4.4	6	10	0.5
222	925	15.6	6.1	5.9	15.4	0.7
196	815	14	5.7	5	13	0.5
349	1455	21	8.3	12	30	1.7
337	1408	21	8.3	13	25	3.8
277	1156	18	7.1	7	24	1.5
269	1121	18	7.1	8	20	3.1
440	1834	31	14.4	18	24	0.8
431	1796	31	14.6	18	20	2.7
396	1655	26	10.8	15	26	1.3
388	1616	27	10.8	16	22	3.1

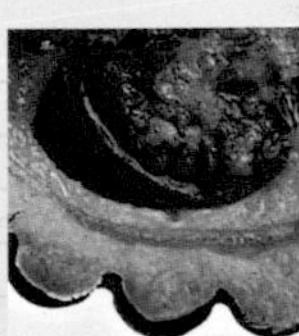

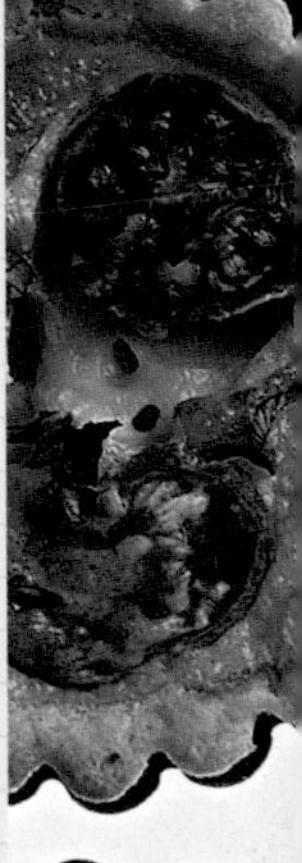

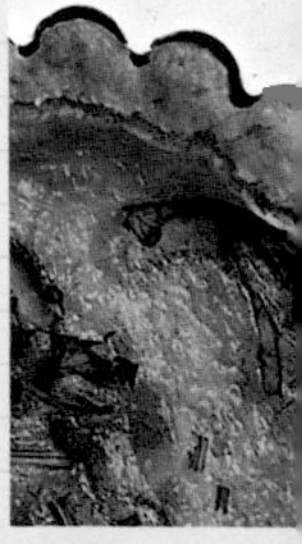

PASTRY DISHES	AVERAGE PORTION g	GI	GL
Quiche, cheese, onion and potato	140	H	H
Quiche, cheese, onion and potato, wholemeal	140	H	H
Quiche, fish, shop bought	140	H	H
Quiche Lorraine	140	H	H
Quiche, meat, shop bought	140	H	H
Quiche, mushroom	140	L	H
Quiche, mushroom, wholemeal	140	L	H
Quiche, spinach	140	L	H
Quiche, spinach, wholemeal	140	L	H
Quiche, vegetable	140	L	H
Quiche, vegetable, shop bought	140	H	H
Quiche, vegetable, wholemeal	140	L	H
Sausage rolls, flaky pastry	60	H	H
Sausage rolls, shortcrust pastry	60	H	H
Steak and kidney pie, double crust	120	H	H
Steak and kidney pie, shop bought	141	H	H

Unless otherwise specified, all pastry dishes are homemade.

ENERGY kcal	ENERGY kJ	FAT g	SATURATED FAT g	PROTEIN g	CARBO-HYDRATE g	FIBRE g
480	2005	33	16	18	28	1.4
472	1966	34	16.1	19	25	3.1
407	1702	25.9	10.9	13.7	31.6	1.5
377	1569	24.6	11.6	12.7	27.6	1.3
474	1985	29.7	11.9	13.6	40.7	1.8
398	1659	27	12.2	14	26	1.3
388	1618	28	12.2	15	21	3.1
287	1203	18	5.6	14	18	2
281	1176	18	5.6	15	16	3.2
295	1238	18	6	7	28	2.1
405	1695	27.6	11.5	11.3	29.7	2.5
286	1197	18	6	8	24	3.9
238	991	17	6.5	6	15	0.8
229	953	16	5.8	6	17	0.9
406	1694	26	9.7	16	27	1.1
437	1826	24	11.8	12	38	0.7

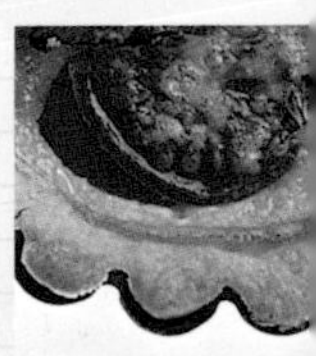
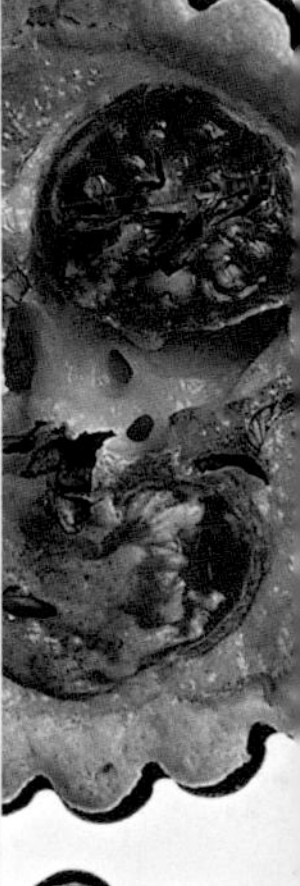
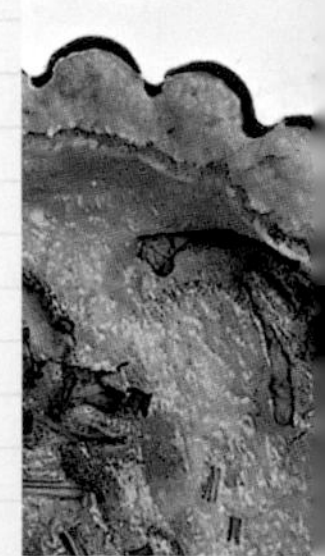

MILK-BASED PUDDINGS AND DESSERTS	AVERAGE PORTION g	GI	GL
Blancmange	150	H	H
Cheesecake, chocolate	120	H	H
Cheesecake, frozen	90	H	H
Cheesecake, fruit, individual	90	H	H
Cheesecake, fruit, large	120	H	H
Fruit fool	120	H	H
Fruit fool, low-fat	120	H	H
Milk pudding, made with semi-skimmed milk	200	H	H
Milk pudding, made with whole milk	200	H	H
Rice desserts, individual, with fruit	135	H	H
Rice pudding, canned	200	H	H
Tiramisu	100	H	H
Trifle, chocolate	113	H	H
Trifle, fruit	113	H	H

OTHER DESSERTS	AVERAGE PORTION g	GI	GL
Bread and butter pudding	100	H	H
Christmas pudding	100	H	H
Crumble, fruit	170	H	H
Jelly, made with water	115	H	H
Mousse, chocolate, individual	60	H	H
Mousse, chocolate	60	H	H
Mousse, fruit	60	H	H
Pavlova, chocolate	100	H	H
Pavlova, fruit	100	H	H
Profiteroles with chocolate sauce	155	H	H
Sponge pudding	100	H	H
Sponge pudding, syrup, canned	100	H	H
Sticky toffee pudding	100	H	H
Strudel, fruit	115	H	H

ENERGY kcal	ENERGY kJ	FAT g	SATURATED FAT g	PROTEIN g	CARBO-HYDRATE g	FIBRE g
171	720	6	3.4	5	27	Trace
415	1731	28	14.4	6	37	0.8
218	915	10	5	5	30	0.8
238	1000	11	6.8	5	31	0.9
353	1477	19	11.8	5	42	1
213	890	13	8.8	3	22	0.5
97	405	6	0	4	8	0.5l
214	914	4	2.2	8	40	0.2
258	1086	9	5.4	8	40	0.2
153	648	3	2	4	29	0.4
178	748	3	3.2	7	28	0.4
275	1149	18.2	10.5	4.8	24.4	0.9
233	971	17	10.7	5	15	1.4
161	672	10	6.3	3	15	2.4
159	667	7.9	4.3	6.2	16.9	0.3
291	1227	10	4.5	5	50	1.3
373	1571	14	6.7	4	61	2.2
70	299	0	0	1	17	0
83	352	4	2.7	2	12	0.1
73	310	2	1.5	3	11	0.1
82	345	3	2	3	11	0.1
370	1552	20	10.7	4	47	0.3
288	1210	13	7.1	3	42	0.3
535	2226	40	21.7	9	38	1.1
343	1437	16.8	3.7	5.7	45.1	1.1
265	1108	9.1	5	3.1	45.4	0.8
304	1274	12.4	7.8	5.7	45.3	1.8
279	1167	16	6	3	33	1.2

PIES	AVERAGE PORTION g	GI	GL
Apple pie, deep filled, double crust	110	H	H
Apple pie, double crust	110	H	H
Banoffee pie	150	H	H
Lemon meringue pie	150	H	H
Mince pies, individual	55	H	H
Mississippi mud pie	150	H	H
Pecan pie	100	H	H
Pumpkin pie	100	H	H
ICE CREAM AND FROZEN DESSERTS			
Arctic roll	50	H	H
Banana split	100	H	H
Chocolate/chocolate mint and nut cone	70	H	H
Chocolate nut sundae	60	H	H
Frozen ice cream dessert, chocolate	60	H	H
Frozen ice cream dessert, plain	60	H	H
Ice cream, chocolate/caramel	80	H	M
Ice cream, fruit	75	H	M
Ice cream, luxury, vanilla	75	H	M
Ice cream, soft scoop, dairy	60	H	M
Ice cream, soft scoop, non-dairy	60	H	M
Ice cream, soya	75	H	M
Ice cream, vanilla	75	H	M
Ice cream, virtually fat-free	75	H	M
Ice cream bar, chocolate-covered	80	H	M
Ice cream bar, chocolate flavoured coating	48	H	M
Ice cream cone, chocolate/mint/nuts	73	H	M
Ice cream cone, strawberry	81	H	M
Knickerbocker glory	100	H	M
Kulfi	80	H	H
Lolly, fruit	73	H	H
Lolly, ice cream, with fruit coating	73	H	H
Peach melba	60	H	H
Sorbet, fruit	75	H	H

ENERGY kcal	ENERGY kJ	FAT g	SATURATED FAT g	PROTEIN g	CARBO-HYDRATE g	FIBRE g
281	1183	12	3.8	4	42	1.1
269	1129	14	4.5	4	33	1.1
478	1997	30	15	6	49	3.8
377	1590	13	4.6	4	65	0.7
233	975	11	4.1	2	32	1.2
478	1997	30	15	6	49	3.8
397	1662	18.5	3.5	4	57.2	3.5
203	852	9.5	1.8	3.9	27.3	2.7
100	424	3	1.5	2	17	Trace
182	761	11	6	2	19	0.6
204	858	10.1	7.7	2.5	27.7	0.8
167	699	9	5	2	21	0.1
150	627	11	8.5	2	13	0
136	568	9	6.7	2	14	Trace
210	880	11.1	7.2	3.4	25.8	0.7
119	499	5	3.4	2	17	0.2
161	670	11	6.8	3	13	0
101	427	4.9	3.1	1.9	13.2	0
115	484	4.6	3	1.6	17.9	0
156	654	9	2.8	2	17	0.1
115	480	6	3.6	2	14	0
76	324	1	Trace	3	15	0
269	1124	16.9	11.4	3.1	27.9	0
142	590	10	8.8	2	11	0
207	867	13	9.6	3	21	0.2
201	844	10	7.1	3	28	0.2
112	473	5	2.9	2	16	0.2
339	1405	32	18.2	4	9	0.5
56	234	1	0.1	0	14	0.1
78	326	2	1	1	15	0.1
98	411	6	3.8	1	10	0.2
98	422	Trace	Trace	1	26	0

CONFECTIONERY	AVERAGE PORTION g	GI	GL
Boiled sweets	7	H	H
Bounty® bar	57	H	H
Chewy sweets	3	H	H
Chocolate, diabetic	50	L	M
Chocolate, dark with crème/mint centres	8	M	M
Chocolate, milk	50	M	H
Chocolate, plain	50	M	M
Chocolate, white	50	M	M
Chocolate-covered bar with fruit/nut	34	M	M
Chocolate-covered bar with wafer/biscuit	34	M	M
Chocolate-covered caramels	5	M	M
Chocolate-covered caramel and biscuit fingers	28	M	M
Chocolate-covered bar with caramel and cereal	30	M	M
Creme egg®	39	H	H
Fruit gums/jellies	2	H	H
Fudge	11	H	H
Liquorice allsorts	7	H	H
Maltesers®	37	H	H
Mars® bar	58	H	H
Marshmallows	5	H	H
Milky Way® bar	52	H	H
Nougat	70	H	H
Peppermint creams	6	H	H
Peppermints	6	H	H
Sherbert sweets	5	H	H
Smartie®-type sweets	15	M	M
Snickers® bar	58	H	H
Toffees, mixed	8	H	H
Truffles, mocha	7	H	H
Truffles, rum	7	H	H
Turkish delight, with nuts	15	H	H
Turkish delight, without nuts	15	H	H

ENERGY kcal	ENERGY kJ	FAT g	SATURATED FAT g	PROTEIN g	CARBO-HYDRATE g	FIBRE g
23	98	Trace	0	Trace	6	0
270	1129	15	12.1	3	33	1.4
11	48	0.2	0.1	0	2.6	0
224	934	15	9.1	5	19	0.6
34	145	1.3	0.8	0.3	5.8	0.
260	1086	15.6	9.4	3.7	28	0.7
255	1069	14	8.4	3	32	1.3
265	1106	15	9.2	4	29	0
170	711	9	4.6	3	20	1.3
170	711	9	4.6	3	20	1.3
25	103	1.2	0.7	0.2	3.5	0
139	582	6.6	3.8	1.4	19.6	0.4
150	631	7.7	4.9	1.6	19.9	0.2
163	681	6	1.8	2	28	0.5
6	28	0	0	0	2	Trace
49	205	2	1	0	9	0
24	104	1	0.3	0	5	0.1
176	739	8.6	5.2	2.8	23.3	0.1
234	988	8.9	4.2	2.4	38.6	0.5
16	70	0	0	0	4	0
231	973	8.2	4	2	39.8	0.3
269	1138	6	0.8	3	54	0.6
22	95	Trace	0	0	6	Trace
24	101	0	0	0	6	0
18	76	0	0	0	5	Trace
68	288	3	1.6	1	11	0.2
278	1163	16.4	5.3	4.4	30.3	1.4
31	130	1.3	0.7	0.2	5	0
34	143	2	1.1	0	4	0.1
36	152	2	1.4	0	3	0.1
52	219	0	0	1	12	0
44	189	0	0	0	12	0

COLD BEVERAGES	AVERAGE PORTION g	GI	GL
Cola	160	H	H
Cola, diet	160	M	M
Ginger ale, dry	160	M	M
Ginger ale, dry, diet	160	M	M
Lemonade	160	H	H
Lemonade, diet	160	M	M
Lucozade	160	H	H
Orangeade	160	H	H
Orangeade, diet	160	M	M
Root beer	160	H	H
Soda, cream	160	H	H
Soda water	160	H	M
Tonic water	160	H	M

FRUIT JUICES			
Acai berry	160	M	M
Apple, fresh	160	M	M
Carrot	160	M	M
Grape, unsweetened	160	M	M
Grapefruit, unsweetened	160	M	M
Lemon, fresh	10	M	L
Lime, fresh	10	M	L
Mango, canned	160	M	M
Orange, fresh	160	M	M
Passion fruit	160	M	M
Pineapple, unsweetened	160	M	M
Pomegranate, fresh	160	M	M
Prune	160	M	M
Smoothie made with fruit or fruit juice	300	M	M
Tomato	160	M	L

ENERGY kcal	ENERGY kJ	FAT g	SATURATED FAT g	PROTEIN g	CARBO-HYDRATE g	FIBRE g
66	278	0	0	Trace	17	0
Trace	Trace	0	0	0	0	0
24	99	0	0	0	6	0
2	6	0	0	0	0.4	0
35	149	0	0	Trace	9	0
Trace	Trace	0	0	0	0	0
96	410	0	0	Trace	26	0
113	474	0	0	0	27	0
1	4	0	0	0	0.4	0
66	275	0	0	0	17	0
82	344	0	0	0	21	0
26	108	0	0	0	6.9	0
53	226	0	0	0	14	0
79	332	0.6	0	0	19.7	0
59	251	0.2	0	0.2	15.5	0
38	165	Trace	Trace	1	9	0.2
74	314	Trace	Trace	0	19	0
53	224	Trace	Trace	1	13	Trace
1	3	Trace	Trace	0	0	0
1	4	Trace	Trace	0	0	0
62	266	Trace	Trace	0	16	Trace
54	234	0.2	0	1	13.6	0.3
75	302	Trace	Trace	1	17	Trace
66	283	Trace	Trace	0	17	Trace
70	302	Trace	Trace	0	19	Trace
91	389	Trace	Trace	1	23	Trace
188	789	0.4	0.1	1.3	47.9	3
22	99	Trace	Trace	1	5	1

FRUIT SQUASHES AND DRINKS	AVERAGE PORTION g	GI	GL
Barley water, orange/lemon flavour	250	H	H
Blackcurrant juice drink	288	H	H
Drink, citrus/apple/mixed fruit flavours	250	H	H
Drink, low-calorie, lemon/orange/ mixed fruit flavours	250	M	M
Drink, low-sugar, fortified with vitamins	250	M	M
Fruit juice drink	206	M	M
Fruit juice drink, carbonated	160	M	M
Fruit juice drink, low-calorie, mixed fruit flavours	206	M	M
High juice drink, orange/lemon flavours	250	M	M

MILK-BASED DRINKS			
Drinking yoghurt	200	M	L
Flavoured milk	214	H	H
Milkshake, with semi-skimmed milk	300	H	H
Milkshake, with skimmed milk	300	H	H
Milkshake, with whole milk	300	H	H
Milkshake syrup, with semi-skimmed milk	290	H	H
Milkshake syrup, with skimmed milk	290	H	H
Milkshake syrup, with whole milk	297	H	H
Smoothie made with fruit juice and milk/yoghurt	300	M	M
Soya milk, flavoured	146	L	L
Probiotic yoghurt drink, orange	100	M	M
Probiotic yoghurt drink, plain	100	L	L

HOT BEVERAGES			
Cappuccino, with semi-skimmed milk	190	M	H
Cappuccino, with skimmed milk	190	M	H
Cappuccino, with whole milk	190	M	H

ENERGY kcal	ENERGY kJ	FAT g	SATURATED FAT g	PROTEIN g	CARBO-HYDRATE g	FIBRE g
35	150	0	0	0	9	0
86	372	0	0	Trace	23	0
48	200	0	0	Trace	13	0
3	8	0	0	Trace	1	0
10	43	0	0	Trace	3	0
76	328	Trace	Trace	0	20	Trace
62	264	Trace	Trace	Trace	16	Trace
21	89	Trace	Trace	0	5	Trace
63	268	0	0	0	16	0
124	526	Trace	Trace	6	26	Trace
146	614	3	1.9	8	23	0
207	882	5	3	10	34	Trace
171	726	1	0.3	10	34	0
261	1104	11	7.2	9	33	Trace
171	725	4	2.6	8	27	0
136	576	1	0.2	9	27	0
223	936	10	6.2	8	27	0
293	1226	4.8	2.8	5.1	61.1	4.5
58	245	2	0.3	4	5	Trace
67	279	0.9	0.6	1.5	13.4	1.3
68	284	1	0.7	1.7	13.1	1.4
46	193	2	1	3	5	0
33	141	0	0.1	3	5	0
65	269	4	2.3	3	5	0

HOT BEVERAGES	AVERAGE PORTION g	GI	GL
Cocoa, with semi-skimmed milk	250	M	M
Cocoa, with skimmed milk	250	M	M
Cocoa, with whole milk	250	M	M
Coffee and chicory essence, with water	190	L	L
Coffee, filter	190	L	L
Coffee, filter, with semi-skimmed milk	190	L	L
Coffee, filter, with single cream	190	L	L
Coffee, filter, with skimmed milk	190	L	L
Coffee, filter, with whole milk	190	L	L
Coffee, instant	190	L	L
Coffee, instant, with semi-skimmed milk	190	L	L
Coffee, instant, with skimmed milk	190	L	L
Coffee, instant, with whole milk	190	L	L
Drinking chocolate, with semi-skimmed milk	190	H	H
Drinking chocolate, with skimmed milk	190	H	H
Drinking chocolate, with whole milk	190	H	H
Latte, with semi-skimmed milk	190	L	L
Latte, with skimmed milk	190	L	L
Latte, with whole milk	190	L	L
Mocha with semi-skimmed milk	190	M	M
Mocha with skimmed milk	190	M	M
Mocha with whole milk	190	M	M
Ovaltine, with semi-skimmed milk	190	H	H
Ovaltine, with skimmed milk	190	H	H
Ovaltine, with whole milk	190	H	H
Tea, black	190	L	L
Tea, Chinese	190	L	L
Tea, green	190	L	L
Tea, herbal	190	L	L
Tea, lemon, instant	190	M	M
Tea, with semi-skimmed milk	190	L	L
Tea, with skimmed milk	190	L	L
Tea, with whole milk	190	L	L

ENERGY kcal	ENERGY kJ	FAT g	SATURATED FAT g	PROTEIN g	CARBO-HYDRATE g	FIBRE g
143	608	5	3	9	18	0.5
110	475	1	0.8	9	18	0.5
190	800	10	6.5	9	17	0.5
17	76	Trace	0	0	5	0
4	15	Trace	Trace	0	1	0
13	55	1	0.2	1	1	0
27	106	2	1.3	1	1	0
11	43	0	0	1	1	0
13	59	1	0.6	1	1	0
Trace	4	Trace	0	0	Trace	0
13	55	1	0.2	1	1	0
8	39	0	0	1	1	0
15	65	1	0.6	1	1	0
135	578	4	2.3	7	21	Trace
112	481	1	0.6	7	21	Trace
171	716	8	4.8	6	20	Trace
60	252	2	1.3	4	7	0
33	141	1	0.1	3	5	0
85	353	5	3	4	6	0
96	405	5	3.1	4	9	0.1
81	339	3	2	4	9	0.1
120	500	7	4.8	4	9	0.1
150	642	3	1.9	7	25	Trace
129	549	1	0.2	7	25	Trace
184	779	7	4.6	7	25	Trace
Trace	4	Trace	Trace	0	Trace	0
2	10	0	0	0	0	0
Trace	Trace	0	0	0	Trace	0
2	25	Trace	Trace	0	0	0
15	65	0	0	0	4	0
13	53	1	0.2	1	1	0
8	36	0	0	1	1	0
15	61	1	0.6	1	1	0

HOT DRINKS

	AVERAGE PORTION g	GI	GL
Beers			
Bitter, best/premium	287	H	H
Bitter, bottled	287	H	H
Bitter, canned	287	H	H
Bitter, draught	287	H	H
Bitter, keg	287	H	H
Bitter, low-alcohol	250	H	H
Brown ale, bottled	250	H	H
Mild, draught	287	H	H
Pale ale, bottled	250	H	H
Strong ale/barley wine	287	H	H
Ciders			
Dry	287	M	M
Low-alcohol	250	M	M
Sweet	287	H	H
Vintage	287	H	H
Lagers			
Bottled	250	H	H
Canned	287	H	H
Draught	287	H	H
Low-alcohol	250	H	H
Premium	287	H	H
Shandy	287	H	H
Liqueurs			
Advocaat	25	M	M
Campari	28	M	M
Cherry Brandy	25	H	H
Coîntreau	25	H	H
Cream liqueurs	25	H	H
Crème de Menthe	25	H	H
Curaçao	25	H	H
Drambuie	25	H	H
Egg nog	160	H	H
Grand Marnier	25	H	H

ENERGY kcal	ENERGY kJ	FAT g	SATURATED FAT g	PROTEIN g	CARBO-HYDRATE g	FIBRE g
95	399	Trace	Trace	1	6	Trace
86	356	Trace	Trace	1	6	Trace
92	379	Trace	Trace	1	7	0
92	379	Trace	Trace	1	7	0
89	370	Trace	Trace	1	7	0
33	135	0	0	1	5	Trace
75	315	Trace	Trace	1	8	Trace
69	293	Trace	Trace	1	5	Trace
70	295	Trace	Trace	1	5	Trace
189	789	Trace	Trace	2	18	Trace
103	436	0	0	Trace	7	0
43	185	0	0	Trace	9	0
121	505	0	0	Trace	12	0
290	1208	0	0	Trace	21	0
73	300	Trace	Trace	1	4	0
83	347	Trace	Trace	1	Trace	Trace
83	347	Trace	Trace	1	Trace	Trace
25	103	Trace	Trace	1	4	Trace
169	700	Trace	Trace	1	7	Trace
32	138	0	0	Trace	9	Trace
65	273	2	0.5	1	7	0
65	272	0	0	0	6.8	0
66	275	0	0	Trace	8	0
79	328	0	0	Trace	6	0
81	338	4	0	Trace	6	0
66	275	0	0	Trace	8	0
78	326	0	0	Trace	7	0
79	328	0	0	Trace	6	0
182	763	7	3.4	6	16	0
79	328	0	0	Trace	6	0

LIQUEURS	AVERAGE PORTION g	GI	GL
Pernod	25	H	H
Southern Comfort	25	H	H
Tia Maria	25	H	H
Fortified wines			
Port	50	M	L
Sherry, dry	50	H	L
Sherry, medium	50	H	L
Sherry, sweet	50	H	L
Tonic wine	125	H	L
Vermouth, dry	48	H	L
Vermouth, sweet	48	H	M
Spirits, 40% volume			
Brandy	25	L	L
Gin	25	L	L
Rum	25	L	L
Vodka	25	L	L
Whisky	25	L	L
Stout			
Bottled	250	H	H
Extra	287	H	H
GuinnessTM	287	H	H
Mackeson	287	H	H
Wines			
Champagne	125	L	L
Mulled wine	125	L	L
Prosecco	100	L	L
Red wine	125	L	L
Rosé, medium	125	L	L
White wine, dry	125	L	L
White wine, medium	125	L	L
White wine, sparkling	125	L	L
White wine, sweet	125	L	L

ENERGY kcal	ENERGY kJ	FAT g	SATURATED FAT g	PROTEIN g	CARBO-HYDRATE g	FIBRE g
79	328	0	0	Trace	6	0
79	328	0	0	Trace	6	0
66	275	0	0	Trace	8	0
79	328	0	0	0	6	0
58	241	0	0	0	1	0
58	241	0	0	0	3	0
68	284	0	0	0	3	0
159	665	0	0	Trace	15	0
52	217	0	0	0	1	0
72	303	0	0	Trace	8	0
52	215	0	0	Trace	Trace	0
52	215	0	0	Trace	Trace	0
52	215	0	0	Trace	Trace	0
52	215	0	0	Trace	Trace	0
52	215	0	0	Trace	Trace	0
93	390	Trace	Trace	1	10	0
112	468	Trace	Trace	1	6	0
86	362	Trace	Trace	1	4	0
103	439	Trace	Trace	1	13	0
95	394	0	0	0	2	0
245	1028	0	0	0	32	0
69	289	0	0	0.1	2	0
85	354	0	0	0	0	0
89	368	0	0	0	3	0
83	344	0	0	0	1	0
93	385	0	0	0	4	0
93	384	0	0	0	6	0
118	493	0	0	0	7	0

Special Photography
© Octopus Publishing Group Limited/Gareth Sambidge.

Other Photography
© Octopus Publishing Group Limited/Ian O'Leary 8–15, 16–33, 80–85, 86–93, 130–131, 132–143.

Commissioning Editor Nicola Hill
Managing Editor Clare Churly
Design Manager Tokiko Morishima
Designer Peter Gerrish
Picture Librarian Sophie Delpech
Production Assistant Nosheen Shan